PSYCHOLOGY RESEARCH PROGRESS

PSYCHE, MATTER AND CROP CIRCLES

AN APPROACH BY ANALYTICAL PSYCHOLOGY

PSYCHOLOGY RESEARCH PROGRESS

Additional books in this series can be found on Nova's website under the Series tab.

Additional e-books in this series can be found on Nova's website under the e-book tab.

PSYCHOLOGY RESEARCH PROGRESS

PSYCHE, MATTER AND CROP CIRCLES

AN APPROACH BY ANALYTICAL PSYCHOLOGY

SIMONE MANCINI CASTILHO

NOVINKA

New York

For permission to use material from this book please contact us:
Telephone 631-231-7269; Fax 631-231-8175
Web Site: http://www.novapublishers.com

Additional color graphics may be available in the e-book version of this book.

Library of Congress Cataloging-in-Publication Data

ISBN: 978-1-63321-587-0

Published by Nova Science Publishers, Inc. † New York

For my family

CONTENTS

PREFACE*

About 2 years ago through an email from a friend about the 'crop circles', I came across pictures of this phenomenon for the first time.

I was hypnotized.

There was in me a mixture of affection and amazement when I realized that those images were formed by the contrast of colors and lights, arising from the position of the plants themselves in the field. The magnitude, the accuracy of lines and geometric perfection present in those aerial photos were, in my view, true works of art.

I was confused, absorbed and impressed. I looked at the pictures for a long time, I tried to understand how the circles appeared, and remember that I just stood there, in unanswered awe, which requested my reverence and, at the same time, offered me no conclusions. I remained in that started and, without realizing it, I kept gestating the images, the colors of the plantation, the holy perfection of the forms and all of my many questions.

Quantum physics for some time had rendered me uncomfortable to take from me the truths that supported me, so peacefully, expanding the relationship between psyche and matter in ways I never imagined. Within the theoretical perspective of Swiss psychiatrist Carl Gustav Jung -namely, analytical psychology-, I have substantiated aspects of psychophysical nature and dared raise hypotheses for the phenomenon of crop circles. I immersed myself in reading; I tried to maintain a critical perspective and made it to the pages of this book, which I now present to the reader.

It is important to remember, however, that we need to see ideas and beliefs not as immovable objects but as vehicles of growth and change. The Hundredth Monkey Story helps us to understand it [1]:

---

* English Version: Graziella Gallo

The Japanese monkey, Macaca Fuscata, had been observed in the wild for a period of over 30 years.

In 1952, on the island of Koshima, scientists were providing monkeys with sweet potatoes dropped in the sand. The monkey liked the taste of the raw sweet potatoes, but they found the dirt unpleasant.

An 18-month-old female named Imo found she could solve the problem by washing the potatoes in a nearby stream. She taught this trick to her mother. Her playmates also learned this new way and they taught their mothers too.

This cultural innovation was gradually picked up by various monkeys before the eyes of the scientists. Between 1952 and 1958 all the young monkeys learned to wash the sandy sweet potatoes to make them more palatable. Only the adults who imitated their children learned this social improvement. Other adults kept eating the dirty sweet potatoes.

Then something startling took place. In the autumn of 1958, a certain number of Koshima monkeys were washing sweet potatoes -- the exact number is not known. Let us suppose that when the sun rose one morning there were 99 monkeys on Koshima Island who had learned to wash their sweet potatoes. Let's further suppose that later that morning, the hundredth monkey learned to wash potatoes.

Then it happened!

By that evening almost everyone in the tribe was washing sweet potatoes before eating them. The added energy of this hundredth monkey somehow created an ideological breakthrough!

But notice: A most surprising thing observed by these scientists was that the habit of washing sweet potatoes then jumped over the sea... Colonies of monkeys on other islands and the mainland troop of monkeys at Takasakiyama began washing their sweet potatoes.

Thus, when a certain critical number achieves an awareness, this new awareness may be communicated from mind to mind.

Although the exact number may vary, this Hundredth Monkey Phenomenon means that when only a limited number of people know of a new way, it may remain the conscious property of these people.

But there is a point at which if only one more person tunes-in to a new awareness, a field is strengthened so that this awareness is picked up by almost everyone.

The story of the Japanese monkeys is a good example of the propagation of a paradigm shift.

I hope that one of those who reads this book turns out to be the one that makes the great difference.

Chapter 1

THE WESTERN WORLDVIEW AND THE CURRENT MODEL OF SCIENCE

"Immersed in a materialistic perspective, matter became our own view. Its concrete objects are our plans, wishes and promises of happiness. We admit our thoughts as a result of neural circuits, and our emotions, a product of neurochemistry.

We research, explore, confidently dive into the depths of matter, into its atom.

And then we find this immense void full of possibilities, of which we are part. Now, it is the void that gazes at us, the void of matter pregnant with meaning, reflecting our psyche.

Will we be ready to see the whole?

(Simone Mancini Castilho)

"And if you gaze long enough into an abyss, the abyss will gaze back into you."

(Friedrich Nietzsche)

The history of modern science is the history of the development of materialistic mechanistic thinking.

This mechanistic paradigm was first formulated and disseminated by the philosophy of Descartes. Cartesian thought had a significant influence on physics of that time, and on the later development of this science by Newton.

For Descartes, God Creator radically transcended nature, not being subjected to any prior truth. By virtue of the power of His free will, he created truths.

It is important to understand that this radical transcendence of God has two fundamental consequences: human free will and independence of science.

According to Descartes, men are not a part of God. The transcendence of the Creator removes any pantheism. Men, simple creatures overtaken by their creator (in this perspective, I conceive God because I find in me the mark of His infinity, although I do not understand Him), are thereby granted full autonomy. Men are free, and they can answer yes or no to God's commands.

God proposes and man, through his free will, puts into practice. This way, God is not to blame for my mistakes or my sins. It is me the one who makes mistakes, it is me who sins. My free will makes me worthy or guilty.

Likewise, the transcendence of God will make possible a purely rational and mechanistic science of nature.

Nature, according to Descartes, is not dynamic by itself. All dynamics belong to the creator. To the extent that nature is stripped of all metaphysical depth, all purpose disappears and nature is reduced to a fully transparent mechanism for mathematical language. Nature has nothing divine, is a created object, located on the same level of human intelligence, and, therefore, entirely dedicated to its exploitation. This is at the same time, the rejection of every pagan naturalism (nature is not a goddess) and the metaphysical foundation of scientific rationalism.

Not everything has the same value in the scientific work of Descartes. If his point of view and his considerations on algebraic expression of curves (he is, along with Fermat, the inventor of analytic geometry) are indisputable scientific contribution, one cannot say the same about his views in the field of physics. But the spirit of Cartesian physics and physiology - treated as a chapter of physics - is nothing more than the spirit of mechanism. When Descartes says that animals are machines, he argues, primarily, that it is possible to explain physiological functions through mechanisms. And this leads to modern science. For Descartes, the physical world has no mysteries. Things determine each other (laws of shock) by direct contact, in a space where emptiness does not exist.

As mentioned previously, despite having been a pioneer of mechanism, Descartes failed to formulate a consistent physical theory capable of explaining natural phenomena. Departing from mechanism, it was Newton who did it.

Newton is the author of *Philosophiae Naturalis Principia Mathematica*, published in 1687, which describes the law of universal gravitation and the so-called Newton's Laws — the three laws of moving bodies that have established themselves as the basis for classical mechanics. By demonstrating the consistency that existed between the system he devised and Kepler's laws of planetary motion, he was the first to demonstrate that the movement of objects, both on Earth and on other celestial bodies are governed by the same set of natural laws. The unifying and predictive power of his laws was centered in scientific revolution, in the advancement of heliocentricism and the widespread notion that rational investigation can reveal the more intrinsic operation of nature.

Our worldview, rooted in the ideas of Descartes and in Newton's discoveries, composing classical physics, is basically causal and deterministic. From this perspective, the world has its own laws and it is up to men to decipher them and thus do science. The credibility of scientific results increases the further experiments avoid interference between observer and the object observed. According to this view, man and the world are independent and separate, as well as psyche and matter. There is an external world outside us, composed of matter or particles that, subjected to a causal agent, responds with a specific and determined effect, within a framework of time and space known to us. Moreover, often the totality of our psyche is admitted as belonging exclusively to our inner world and matched to our restricted conscious sphere. We believe without question that our reason, will and as little interference as possible in the object studied is absolutely feasible and, moreover, the ideal way of doing science.

Newtonian science gives us concepts that are still quite entrenched, such as materialism, objectivity and determinism that are only appropriate and justified when we investigate the order of the macroscopic external world.

In classical Newtonian physics, atoms were seen as solid objects (nucleus) surrounded by particles of electrons, hence the basis of all materialistic thought.

Another key idea is that in which all objects are independent and separate from the mind, allowing us to objectively perceive the world as it really is, without any interference from the observer.

In Cartesian-Newtonian view of the world, the universe can be thought of as a large number of billiard balls - large and small - in a three-dimensional table, which we call space. If we knew, on all occasions, all the forces that act on each of these balls, that is, if we know its baseline conditions - positions and velocities at some initial time - we would be able to calculate where each of these bodies would be found in all future occasions.

These are two fundamental principles of classical physics: objectivity and determinism.

The third principle was discovered by Albert Einstein. Einstein's theory of relativity, an extension of classical physics to bodies moving at high speed, required the highest speed on Nature's roads to be the speed of light. This speed is huge - 300,000 kilometers per second - but still limited. The implication of this speed limit is that all the influences between material objects that are felt in space-time should be local: they must travel through space a little at a time, with a finite velocity. This is called the principle of locality.

Within this perspective and paradigm, psyche and matter are clearly divided and separate, and matter is still seen as dominant in relation to the psyche. Until quite recently, psyche did not even exist as reality. Carl Gustav Jung was the one responsible for showing the autonomy and relevance of what he called psychic reality, forcing us to review and revise the boundaries, until then clear and independent, between material and psychological domain, as discussed below.

This is the Western worldview and the current model of science that is slowly being questioned and needs to be rethought.

In the words of Jung:

> A view of the world is a hypothesis and not an article of faith. The world changes its face – *tempora mutantur et nos in illis* (times change, and we change with them) for we can only grasp the world as a psychic image in ourselves, and it is not always easy to decide, when the image chances, whether the world or ourselves have changed, or both. The picture of the world can change at any time, just as our conception of ourselves changes. Every new discovery, every new thought, can put a new face on the world. We must be prepared for this, or else we suddenly find ourselves in an antiquated world, itself a relic of lower levels of consciousness [2].

If we are to observe and perhaps adjust our views regarding our conceptions of the relationship between inside and outside, or between psyche and matter, we must present a radically new conception of matter, different from that of quantum mechanics, different from the materialism adopted by classical physics or our unreflected ideas about matter [3].

Similarly, the concept of psyche proposed by Jung and the phenomenon of synchronicity also bring new insights on the boundaries of Psyche, being crucial to this proposition of rethinking our current scientific model.

Last, another phenomenon called crop circle, whose origin has been the source of much controversy, deserves to be addressed so that we can gain a new insight into the connections between psyche and matter, still unusual to our current view of the world.

Chapter 2

QUANTUM PHYSICS

"Anyone not shocked by quantum mechanics has not yet understood it."

(Niels Bohr)

Quantum mechanics is the theory that describes the behavior of matter on the scale of "very small;" that is, it is physics of the components of matter: molecules, atoms and nuclei, which in turn are composed of elementary particles.

From the theoretical point of view, the concept that defines an elementary particle is, above all, of an abstract and mathematical nature. All elementary particles are described by mathematical objects called wave functions from which information about the dynamics of such particles are extracted. The wave function that describes an elementary particle is not reducible to the wave function of other particles.

The model that includes all elementary particles is called the Standard Model. This model states that all known matter is composed of three types of elementary particles: leptons, quarks and mediators, and that in all 61 particles are enough to build all matter found in the Universe.

Quantum physics involves concepts such as wave-particle duality, also called wave-corpuscle duality or matter-energy duality, which involves a basic property of physical entities in subatomic dimensions that behave or have properties of both particles and waves.

Until the late nineteenth century, the movement of particles was analyzed according to Newtonian mechanics, while radiation of electromagnetic waves was described using Maxwell's equations.

What occurred in the first quarter of the twentieth century was that a particular set of experiments showed conflicting results with this distinction that was made between the behavior of waves and particles.

A series of discoveries, such as energy *quanta* described by Max Planck and the wave-particle duality, proposed for the nature of light as well as of matter, pointed out revealing contradictions, demonstrating that the behavior of both did not seem to be so different from each other.

Planck basically considered that the electrons absorb or emit energy in certain quantities and discontinuously separated (as if they were packages), which he called quanta (later called photons). This explains why it is possible to get tanned in the sun, but not in front of a campfire. For an electron to perform a large quantum leap, with emission of ultraviolet light, a more powerful source of energy than that of burning wood is required.

The quantum leap is indeed a ghost for quantum physics. The electron simply disappears from one level and appears in another (quantum discontinuity). Moreover, it is not possible to know when a particular electron will take a leap, or where to, above a minimum degree of energy. So there is no longer any certainty, and one can only speak of probability. Also, to say that light exists as quanta, as photons, is the same as saying that it is composed of particles - tiny grains of sand. As if these paradoxes about light were not enough, there were questions regarding matter: could a material object such as an electron be both wave and particle –in other words, have duality–, like light does? In 1924, French physicist Louis-Victor de Broglie showed that electrons had both wavelike and corpuscular characteristics, behaving in one way or another depending on the specific experiment.

So, the wave-particle duality has become, as said above, a basic property of quantum mechanics.

It was up to Erwin Schrödinger to devise a wave equation for matter, known as Schrödinger's equation, a fundamental part of mathematics that replaced Newton's laws in the new physics.

Electron waves are, nonetheless, probability waves; that is, they relate to where we have more probabilities of finding the particle, said physicist Max Born, who postulated that electrons surrounding the atomic nucleus exist in orbits with quantized energy levels. If there is a great probability of finding the particle, the amplitude of the wave will be wider.

Nevertheless, probability leads to uncertainty. Heisenberg's principle deals precisely with this uncertainty when he says that we cannot simultaneously determine the precise position and velocity (or momentum) of an electron; the slightest effort to measure exactly one of them makes our knowledge of the other vague. Thus, even the Bohr orbits do not provide an accurate description of the whereabouts of an electron; the position of the real orbit is vague. Furthermore, over time, the wavelike state of the packet of electrons would tend to spreading and detachment of the core. Here comes a very important point: the act of observing and measuring the presence and, dare I say, the psyche of the observer modifies the experiment and should be included in the calculations. *The act of measuring reduces the wavelike electron to a particle state.* And such collapse is discontinuous, which means it does not require time and occurs in unpredictable places. Observing electrons, said physicist-philosopher Henry Margenau, is like watching fireflies on a summer night. We can see a glimpse here and a flash of light there, but we have no idea where the firefly is between each of these points. We cannot, with any confidence, define its trajectory.

Last, Bohr's principle of complementarity establishes that electrons are not waves or particles because their true nature transcends both descriptions.

Similarly, subject and object are inextricably mixed [4].

In fact, quantum physics gives us a broader view, casting the possibility of expanding our consciousness to this duality that is being presented.

It all indicates that quantum mechanics is the correct theory to describe physical phenomena at any energy scale. The macroscopic universe would only be a particular case to which there is a more efficient way to describe it, the Newtonian mechanics. Thus, Newtonian physics can be understood as a particular case of quantum mechanics, but the reverse is not true.

Many authors, feeling uncomfortable with the so called orthodox interpretation, try to create alternative theories to replace it. However, it should be noted that, despite its strangeness, quantum mechanics did not present any fault since it was developed in the 1920s. In this way, we have no experimental evidence pointing towards where to find issues capable of dismissing it.

An Indian tale:
The Mouth of the Child

This classic tale took different forms in India. Here is one:

When Krishna (which in Indian mythology is the eighth incarnation of Lord Vishnu, the beneficent force that preserves the worlds) was still a child, he was known for his naughtiness and gluttony. At least, that's what later accounts of his life say.

He wanted milk, jams and sweets. One day, he was accused by his peers to have swallowed a bit of mud, something that is not done. He was severely reprimanded by his earthly mother, Yashoda, who explained to him that never, in no event, he should eat mud.

- I have not eaten mud - told her the very young Krishna
-Okay, open your mouth and show me - said the mother.

The boy obeyed and opened his mouth. And then, leaning over, Yashoda suddenly saw trees, rivers, mountains; she saw the Universe extending in the distance, full of stars and lights; she saw the simultaneous existence of all living beings; she even saw the past and future, the dead that seemed alive and those who were not yet born, but already had a visible form. She also saw all the emotions that living beings can feel: fear, anger and astonishment; she saw laughter and tears, and the three states of matter.

Nothing escaped her. She saw all the things in the mouth of the child. And every part of the Universe was in its place.

- All right - she said, leaning down slightly. - You can close your mouth.

Contos Filosóficos Do Mundo Inteiro

(Jean-Claude Carriere) [5]

"We never doubted that the Universe contains us. Now, we must be able to deal with the mystery of having the Universe within ourselves."

(Simone Mancini Castilho)

2.1. Comparing Classical Physics and Quantum Physics

For clarity, we need to compare classical physics and quantum physics according to certain principles:

- Objectivity
- Principle of causality
- Principle of locality

Objectivity

The current materialistic science implies a material universe that is objective, independent and separate from us, this being a prerequisite for our science to be considered serious.

Quantum physics teaches us that the outcome of the experiment depends on the observer. Therefore, subject and object are inextricably connected so that the validity of the objectivity of classical physics ends up seriously questioned.

Principle of Causality (Determinism)

Classical science is also based on the assumption that if we know the initial or causal conditions that act on an object, we are able to determine the effect of its behavior.

Quantum physics shows us, through the principle of uncertainty, that it is not possible to determine simultaneously and accurately some parameters of quantum objects, whose behavior is essentially probabilistic.

Principle of Locality

This principle of classical physics is based on the assumption that all interactions between objects, seen from this point of view as material and independent, occur through signals that travel through space.

Quantum physics shows us, however, that the influence of the measurement is instantaneous and does not travel locally, which fatally contradicts the idea of locality.

2.2. Quantum Physics and New Perspectives

The discoveries of quantum physics can no longer be denied. The more subatomic particles are observed, the more evident is the fact that classical physics, without losing its validity, is restricted to the macroscopic world. What quantum dimension proposes as new impacts our worldview to the point of us having difficulty understanding or accepting it. On a microscopic level, we do not observe a world of material objects that behave in a consistent and defined manner after undergoing a causal action. Unobserved, the electron is not a particle or a wave, but a function of probability, and determining its position and speed simultaneously cannot be done with certainty. The significance of this is that, when not being measured, the quantum object is both particle and wave, and spreads itself existing in more than one place simultaneously as a cloud of possibility. In fact, electrons are not waves or particles because their true nature transcends both descriptions. These findings confuse us because we experience our macroscopic world as an independent, localized, material and determined reality. But we know that the macroscopic is ultimately composed of the microscopic, obeying its unique laws, which are presented here.

The question to be asked is: Why don't we observe these phenomena in our daily lives, that is, on macroscopic objects? There are two reasons for this. The first is that due to Planck's constant having a very small value, macroscopic objects will have a wavelength well below the limit of detection. Therefore, the wave-particle duality is not evident in our everyday experience, but it is very important for electrons, atoms and molecules.

Second, there is the so called effect of decoherence. This effect has only recently started to be studied and has to do with the fact that we cannot separate a macroscopic body from the environment where it is. Thus, the medium will have a decisive influence on the dynamics of the system, making the necessary conditions to maintain quantum effects disappear on a very short time scale.

In fact, the ever-so-obvious nature of reality has never proved so controversial and difficult to conceptualize as now. The notion of connectivity of the Universe, derived from the principle of non-locality, and the question of subject and object being inseparably connected to each other make this struggle evident. It also reveals the impossibility of having unrestricted access to reality, which makes the possible concepts associated to its nature inevitably relative and partial [6].

To sum up, we need reconciliation and comprehensiveness. In fact, quantum physics forces us to revise our paradigms, not necessarily aiming to replace them but in fact to expand them, providing a wider perspective and a new context that broadens our perception, leading it to a new domain.

Chapter 3

ANALYTICAL PSYCHOLOGY

"All the same, every science is a function of the psyche, and all knowledge is rooted in it. The psyche is the greatest of all cosmic wonders and the sine qua non of the world as an object"

C. G. Jung

During the twentieth century, along with the discoveries of physics, the development of psychology introduces the psyche in the context of scientific studies. At first, the concept of the unconscious mind created by Freud as the region of the psyche with content that the individual is unaware of and formed by suppressed personal tendencies innovates and expands the way in which the psyche was seen, regarded until then as consisting solely of our conscious perceptions.

Based on his clinical experience, Freud believed that the source of emotional disorders resided in repressed traumatic experiences early in life. Thus, he assumed that unconscious contents were available to consciousness in a disguised form (through dreams and slips of the tongue, for example). In this sense, Freud developed psychoanalysis, a therapeutic approach that was intended to make people aware of their own unconscious emotional conflicts.

Thus, the core of the Freud's psychoanalytic approach is the conviction of the existence of the unconscious as:

- A receptacle of traumatic repressed memories;
- A reservoir of impulses that are a source of anxiety, because they are socially or ethically unacceptable to the individual.

Carl Gustav Jung was born in Switzerland in 1875 and died in 1961, at the age of 85 years, being recognized as one of the leading thinkers of the twentieth century.

In 1900, he graduated with honors in medicine and began working as an assistant at the Burgholzli Hospital in Zurich with famous psychiatrist Eugen Bleuler, who invited him to join his research team. Jung quickly excelled in psychiatry and began writing scientific articles of great impact. In one of these articles, he declares his admiration for some of the propositions of Sigmund Freud, who at the time was being heavily criticized by the medical community. This support proved to be decisive for him to build a closer connection with Freud. They became friends, which led to 357 letters, trips, mutual assessment dreams and family gatherings. It is noteworthy that Jung was a defender of the theories of Freud, but did his own research at the Burghőlzgli.

In 1910 when the "International Psychoanalytical Association" was founded, Jung was elected its chairman. In 1914, Jung resigned the presidency of the Association and is no longer a member due to theoretical divergence between his thinking and Freud's. Jung took into account the essential value of sexuality as a fundamental expression of psychological wholeness, but unlike Freud, he did not consider it as the only one. Jung was concerned with deepening sexuality beyond the reach of biological function and personal meaning, addressing its spiritual side and numinous sense. Jung enters a period of personal crisis but allows himself to explore it researching his own unconscious mind. This phase derived the basic material on which he organized his personality psychology, laying the foundations of the Analytical Psychology.

Unlike Psychoanalysis, Jung's Analytical Psychology also expands the concept of psyche, for, in addition to being constituted by the conscious and personal unconscious, the psyche would also present a deeper unconscious layer, a collective mental stratum that Jung called the collective unconscious or objective psyche.

The collective unconscious is concerned with broad structures of the human mind that belong to all human beings and not to one or another individual. Analyzing the dreams and fantasies of his patients, Jung conceived their content as a combination of universally predominant patterns and forces, said archetypes and instincts. In his view, there is nothing individual or singular in human beings at this level; we all have the same archetypes and instincts, and individually should be sought in other areas of one's personality. In The Archetypes and the Collective Unconscious Jung explains [6]:

> A more or less superficial layer of the unconscious is undoubtedly personal. We call it the *personal unconscious.* This, however, rests upon a deeper layer, which no longer has its origin in experiences or personal acquisitions, being innate. This deeper layer is what we call the *collective unconscious.* I choose the term "collective" because the unconscious is not individual but universal in nature.

The collective unconscious would be formed by archetypes, construed as innate dispositions or unobservable structures that are not limited to the psyche, and which, in typical recurring situations, would produce ideas, thoughts, emotions and fantasy ideas, similar to all cultures and all peoples [8].

In other words, archetypes are inherited possibilities to represent similar images; they are instinctive ways of imagining. They are archaic patterns where analogous or likewise configurations are formed. Jung compares archetypes with the axial system of crystals. It determines the crystalline structure in a saturated solution, but does not exist in itself. [...] Whatever its origin, the archetype functions as a node of concentration of psychic energy. When this potential energy state is updated, it takes shape and then we have the archetypal image. We cannot name the image as an archetype because the archetype is merely virtuality. [...] The concept of archetype, postulating the existence of a common psychological basis to all human beings, allows us to understand why similar themes appear in distant places and times in fairy tales, myths, religious dogmas and rites, in the arts, and in philosophy; that is, in the production of the unconscious, generally speaking - whether in the dreams of ordinary people, or in the delusions of madmen [9].

3.1. Comparing the Personal Unconscious and the Collective Unconscious

The personal unconscious, consisting of repressed contents of the history of each individual, somehow withholds the idea of independence and separation between what happens internally and is considered subjective, and what is characterized as external reality, thus strengthening the paradigm of objectivity of the classical science.

Comparatively, the concept of the collective unconscious or objective psyche proposed by Jung's analytical psychology expands the exclusively individual character of psyche, providing a common psychological basis to all individuals formed by archetypes.

Another element that emerges from the collective unconscious and the archetypes that compose it is even more challenging to the prevailing notion of the psyche located within the person and the world, separate and outside him. The point is that, given the regular and universal character of the archetype, doubt is cast on its psychic exclusivity. In Jung's words:

> The fleeting, arbitrary, nebulous and unique character that the mind of the lay always associates to the idea of psychic, only adapts to consciousness, and not to the unconscious. Therefore, the units with which the unconscious operates are definable, not quantitatively but qualitatively; in other words, archetypes thus have a nature about which *we cannot say for sure that is purely psychic* [10].

> The concepts of collective unconscious and archetype, whose nature is not limited to the psychic, unlike the personal unconscious, forces us to expand the way we understand reality. We now need, as in the case of quantum physics, paradigms that allow complementarity and wholeness. Only then we will be able to see phenomena in which psyche and matter insist on walking hand in hand.

3.2. Analytical Psychology and New Perspectives

Furthermore, we can no longer ignore the consequences of the extent to which the psyche is understood within analytical psychology. The collective perspective of the unconscious cannot be fully understood within the current scientific materialist model that internalizes the psyche and makes it exclusive to the individual.

Completely unconscious aspects of the psyche and which remain unknown due to their very nature -called by Jung as of psychoid nature-, point towards a domain where the psychic element seems to blend with the physical world, proposing a type of connection that does not fit in our current model of science and, therefore, troubles us [11].

The only way to honestly absorb the point of view of the analytical psychology is to move towards a new paradigm, also supported by quantum physics, which breaks with the principle of objectivity and realizes that subject and object, psyche and matter are inextricably linked, and may also be understood as different perspectives of a single and unique reality, to be discussed later on.

A zen tale:

Dreaming

The great Taoist master Chuang Tzu once dreamt that he was a butterfly fluttering here and there. In the dream he had no awareness of his individuality as a person. He was only a butterfly. Suddenly, he awoke and found himself laying there, a person once again. But then he thought to himself, "Was I before a man who dreamt about being a butterfly, or am I now a butterfly who dreams about being a man?" [12]

Chapter 4

Analogies between Quantum Physics and Analytical Psychology

"Everything we call real is made of things that cannot be regarded as real"

Niels Bohr

"Without consciousness there would, practically speaking, be no world, for the world exists for us only in so far as it is consciously reflected by a psyche. Consciousness is a precondition of being. Thus the psyche is endowed with the dignity of a cosmic principle, which philosophically and in fact gives it a position co-equal with the principle of physical being."

C. G. Jung

The currents of research in physics and in analytical psychology have converged without the existence of an exchange of opinions. Similar hypotheses were created by the imposition of the objects of study of the two sciences: the atomic and subatomic world in physics and the objective psyche in analytical psychology.

Here are some analogies between quantum physics and analytical psychology:

1. Today physicists are trying to understand matter as transient configurations of energy instead of trying to find its "building blocks". In the same way, the archetypes in the psyche seem to be dynamic systems of psyche energy, rather than static structures [13].

2. The paradox of wave-particle duality was introduced in physics when the wave nature of light was questioned, suggesting that it also exists as a quantum - a separate amount of energy - called photon; and the electron, previously regarded as a particle, would also have a wave nature. Niels Bohr, Danish physicist, described a new way to study this paradox. The idea of Bohr was that the use of a corpuscular or wave box would depend on the experiment in question. In a particular experiment, the phenomenon would be corpuscular or wavelike, never both at the same time. If a phenomenon is represented in a wavelike frame, it cannot be adequately represented in a corpuscular matrix, and vice versa. Bohr's principle of complementarity proposes that such aspects are complementary and not dualistic, given that the true nature would transcend both descriptions. We can notice a parallel in analytical psychology with regard to the unconscious-conscious complementarity. Jung recognized that the relationship between the conscious and the unconscious also forms a pair of complementing opposites. In addition, any attempt to determine the unconscious state makes it change, similarly to what occurs with quantum measurement in atomic physics.

3. Physics also shows us that waves spread continuously through space, while particles have a restricted location [14]. Likewise, Jung tells us that "the area of the unconscious is huge and permanently continuous, while the area of consciousness is the restricted field of a momentary vision" [15].

The Analogies Go Further

4. In quantum theory the notion of causality becomes quite relativized and is reduced to a statistical probability. Some phenomena, as the law of half-life in radioactive decay, cannot be explained by an appropriate cause, but seem rather to manifest a global, a priori order of the cosmic totality. Jung, for his part, was led to postulate an ultimate unity of the objective psyche and to propose the existence of a global order of this unity. Even though relatively separable archetypal structures exist, they are all in an overlapping state and are interdependent. Thus they form a single total acausal order [16].

5. The quantum wave function represents probabilities, in contrast to the collapsed particle. Similarly, the archetypal structures of the unconscious represent the key potentials of psychic manifestation, while the conscious contents are realizations of these potentials [17].

6. Still in the scope of these analogies, physicist Victor Mansfield pointed out a similarity in the way potentials are transformed in reality within both domains [18]:

In physics, the irreversible process of measurement is what turns potential into reality. In psychology, it is reflective consciousness, knowledge through ego, that makes the empirical world possible and brings the transcendent to the multiplicity of it.

7. According to modern physics, four fundamental forces can be distinguished in the universe. Jung observed that our field of consciousness is based on four fundamental functions: thought, intuition, sensation and feeling. Moreover, the mandalas, which are symbols of the psychic totality, have a quaternary structure [19].

For physicist Wolfgang Pauli, who exchanged letters with Jung for 26 years, quantum mechanics corresponds to the psychology of individual processes and of the unconscious in general: the atom, composed of nucleus and its electron cloud, corresponded to the human personality, formed by a "nucleus" (Self) and the ego. Thus, Pauli said that the atom seemed to behave as the archetype of human individual [20]. Jung wrote to him on March 7, 1953, saying that:

> "[...] the lowest mass of a particle consists of particle and wave, in the same manner that the archetype (as a structural element of the unconscious) consists of the static form on the one hand, and the dynamic form on the other." Also in this letter, Jung says:
>
> [...] as a participant medium of Spirit and Matter, I am convinced that the psyche is partly formed of natural matter. Archetypes, for example, are Ideas (in the Platonic sense) on the one hand, and directly connected to physiological processes on the other; and in cases of synchronicity, they are arrangements of physical circumstances, so that they can also be regarded as a characteristic of Matter (as the characteristic that endows it with meaning) [21].

Zen tale: Mind in Motion

Two men were arguing about a banner that fluttered in the wind:

-It is the wind that is really moving! – said the first.

- No, obviously, the banner is moving! – replied the second.

A Zen master, who by chance passed by, heard the discussion and interrupted them, saying:

- It is neither the flag nor the wind that moves; it is the mind that moves.

(Magia Oriental – Autor Desconhecido) [22]

Chapter 5

THE SCIENTIFIC PARADIGMS

Returning to the archetypes that make up the collective unconscious of analytical psychology, it is important to repeat that they are completely unobservable structures that, when stimulated, produce archetypal images and fantasies, intuitions, emotions and thoughts, which are in turn recognizable by us. The archetypal structures or arrangements are inherited, which does not happen with images, updated by each individual.

It must be said that archetypes do not only produce mythic and religious conceptions, similar worldwide, or only basic categories of human fantasy; they also give rise to science in their intellectual premises. Thus, the foundations of all our knowledge, or so-called scientific paradigms, are structurally predetermined by the archetypes in the collective unconscious [23].

We note, therefore, that some ideas in western physics resemble images of Greek philosophy, such as the idea that the entire cosmos would consist of the same raw material or the idea that there would be conservation of this universal matter, as well as a belief in the possibility of transformation of one material substance into another. Another very important notion, already anticipated by Pythagoras and which now stands as a philosophical model in modern physics, is the idea that, ultimately, we deal with mathematical structures. The fact of the responses obtained in the experiments of microphysics being influenced by the observer's presence has forced physicists to turn their attention to their own inner and psychological processes, leading to the concept that knowledge of physics would consist of images that had been purified as mathematical structures [24].

Nowadays, a new vision of reality is based on the understanding of the essential and reciprocal relationships and interdependencies of many phenomena: physical, biological, psychological, social and cultural.

However, can a science that is not based exclusively or preponderantly on measurement exist?, can an understanding of reality that includes the quality and the relationships of things perceived as the experience of them, and, despite this, be called scientific? This understanding is possible. Science does not have to be strictly limited to measurements and quantitative analysis. We must consider any manner of approaching knowledge *scientific* as long as it satisfies two conditions: the empirical base and the process of the creation of models, the two essential elements of the scientific method. Other aspects, like quantification, empirical proof or the use of mathematics, tend to be desirable, but are neither crucial nor indispensable.

Einstein himself used to say that "science consists of creating theories", in other words, in creating representative and coherent models of observed realities. In effect, the process of the creation of models consists of the formation of a logical and coherent network of concepts that are used to link together the observed data. In classical science, these data amounted to quantities obtained through measurement, and conceptual models were expressed, whenever the possibility presented itself, in mathematical language. But now a new systemic precision has appeared, that is much more important, and an objectivity that includes the subject, that represents the whole of reality more integrally [25]

At this point, we need to pay attention to the difficulties and limitations of the experimental method of psychology that describes psychic contents using its own instrument, the psyche. In other words, psychic processes are observed and evaluated by the psychism itself. Physics observes the physical processes in the mental environment; that is, it reconstructs the matter of the physical world, in the psyche. Psychology does not have a neutral ground, external to the psyche itself. As a matter of fact, this is a limiting factor in psychology, as much as the uncertainty caused by the influence of the psyche in physics experiments. It is at this point, however, that the two sciences approach and a question arises: is it possible that matter would mirror psyche? In the eastern world this response is positive and understanding of physical events is a synchronistic reflection of what is happening in the psychic sphere. This brings us to study the boundaries of the archetypal realm where I intend to address synchronicity more deeply and the phenomena of crop circles [26].

Chapter 6

ARCHETYPES, ARCHETYPE BOUNDARIES AND SYNCHRONICITY

"The only experience that we concretely have is our own personal life. The only completed category of our thinking, our professors of philosophy tell us, is the category of personality, every other category being one of the abstract elements of that. And this systematic denial on science's part of personality as a condition of events, this rigorous belief that in its own essential and innermost nature our world is a strictly impersonal world, may, conceivably, as the whirligig of time goes round, prove to be the very defect that our descendants will be most surprised at in our boasted science, the omission that to their eyes will most tend to make it look perspectiveless and short."

William James

6.1. ARCHETYPES

The concept of archetype was modified a few times by Jung throughout his work, and 1912, 1919 and 1946 were remarkable years for its development. The first designation of "archetype" appeared as "primary image" or "primordial image" in his book *Transformations and Symbols of the Libido*, in 1912.

Jung uses the term "archetype" for the first time in 1919, in the essay "Instinct and the Unconscious," presented at the symposium of the University of London and appearing in Volume VIII of his *Collected Works*. In this essay, Jung defines the collective unconscious as being composed of instincts and its correlates, the archetypes. Instincts would be pulses intended to produce actions that result from an inner need, without conscious motivation. The archetypes of perception and apprehension would be the *a priori* forms, innate and intuitive, which are mandatory determinants and inferred to all psychic processes [27].

In 1946, the archetype is defined as a structure that, per se, is completely unobservable but which, when activated, produces archetypal images, ideas and representations. By then, the archetype becomes unrepresentable and the collective unconscious acquires an even more abstract dimension characterized as psychoid. In the words of Jung [28]: “It seems likely to me that the true nature of the archetype is incapable of becoming conscious; that is, it is transcendent, and that is the reason why I call it psychoid.”

6.2. Archetype Boundaries

Jung associates archetype and instinct because, for him, the affinity between the two is complete. The archetype is the image of instinct. At the same time, the archetype is the formative principle of instinctive strength, which is in fact its paradoxical nature [29].

Archetypes are the other face of instinct and are characterized by being innate forms of intuition, universally inherited. If instincts are the impulse that leads to certain action, archetypes are the apprehension of the meaning of action [30]. Nevertheless, we do not experience the instincts or archetypes themselves, given the transcendent and unrepresentable nature of both. We can only access its manifestations, which can be reached through the psyche [31]. The instinctive manifestations experienced as physiological dynamisms or impulses, are familiar to us and we can identify them without difficulty. The archetypal manifestations, on the other hand, have a numinous character and charm and are called by Jung "spiritual". Both are manifestations and form an inseparable pair of opposites [32].

In order to explain the nature of the archetype, Jung makes an analogy of the psyche with the spectrum of light. According to him, experience shows that we use blue, the color of the air and the sky, for spiritual contents, while red, a hot color, expresses feelings and emotions, thus characterizing the color of instinct. Being the archetype a formative principle of instinctive force, as said before, blue is mixed with red, being better represented by violet. Using that same analogy, Jung locates the dynamism of instinct in the infrared part of the spectrum, while the instinctive image would be in the ultra-violet range [33].

Moreover, just as the psychic infra-red, or the biological instinctive psyche, gradually resolves into the physical and chemical processes of the body and of its unfathomable matter, so does the psychic ultra-violet, the archetype, denote a field that cannot be regarded as psychic, even though expressing itself psychically. Its position would be located beyond the limits of the psychic sphere, with its psychoid nature constituting a crossing bridge to matter in general [34].

As already pointed out, despite the archetype *per se* being unrepresentable and its essence being beyond the psyche, we can experience it psychically in the form of ideas and images. Its symbolic manifestations can reach consciousness, whether originating from inside the psychic matrix, or the world around us, or both simultaneously.

According to Aufranc [35]:

> Archetypes, by means of symbols, express themselves in the psychic polarity as well as the biological polarity. The symbol can manifest itself in the psychic polarity of the individual through dream images, fantasies, projections, transfers or psychic symptoms. Collectively, symbolic expression occurs in myths, legends, religions or art. Biological polarity, in turn, has bodily experiences or physical symptoms as a symbolic vehicle.

As we can see, despite keeping the name "archetype" in its most mature and elaborate definition, Jung added an unfolding of subjectivity and objectivity, or, respectively, a psychic facet and another transcendent, unrepresentable, psychoid facet [36].

This dual character is also displayed when Jung describes the property of transgression in archetypes:

> Although they are associated with causal processes or derived from them, [archetypes] are, however, continually surpassing their own limits, a procedure that I would name transgression because *archetypes are not in a definite and exclusive way in the psychic sphere* but can also occur in non-psychic circumstances (equivalence between an external physical process and a psychic process) [37].

Thus, we see the manifestation of the archetypal symbol in the external physical world when, simultaneously to a psychic event, we witness a synchronistic phenomenon. Likewise, the following question could be proposed: Would we be facing the expression of the archetypal symbol in nature's organic matter when we come across images filled with meaning such as the ones presented by the crop circles? These aspects will be discussed later.

> "We live in a hurried time.
>
> We seek to advance, to progress ... We chase efficiency and quick responses. We explain more than we experience. We stand on a timeline, past, present, future, beginning, middle and end, hostages of causes and effects. Connected through virtual networks, yet we lost the connection with our own souls. Globalized, we remain alone. Shorter distances, emotional abysses.
>
> Suddenly, synchronicity captures us. That coincidence that raises suspicion, as improbable as it is, but meaningful and personal to the point that it cannot be ignored. A stumble of time ... which makes us grasp the whole, restoring us, in a glimpse, the chance of feeling included once again. As a part of the whole, and the eternal order from which we never left.
>
> *(Simone Mancini Castilho)*

6.3. Synchronicity

Jung defined synchronicity as the relatively simultaneous occurrence between a psychic event and an event in the external world, whose connection is not by chance, but rather through an equivalence of meaning [38].

A well-known example of synchronicity is the true story of the French writer Émile Deschamps_who in 1805 was treated to some plum pudding by the stranger Monsieur de Fortgibu. Ten years later, he encountered plum pudding on the menu of a Paris restaurant, and wanted to order some, but the waiter told him the last dish had already been served to another customer, who turned out to be de Fortgibu.

Many years later, in 1832, Émile Deschamps was at a diner, and was once again offered plum pudding. He recalled the earlier incident and told his friends that only de Fortgibu was missing to make the setting complete — and in the same instant the now senile de Fortgibu entered the room [39].

Another example:

A mother is working at preparing her dinner, and thinks "It would be nice to have some flowers here today," while her son is in the garden picking flowers for her dinner. The mother has never had flowers on the dinner table before, and the son has never brought flowers, but their close relationship leads them to have the same idea at the same time.

Simultaneous discovery, the creation of the same new idea at causally disconnected places by two persons at approximately the same time. It is very difficult to account for simultaneous discovery by random chance. If for example an American and a British musician, having never had anything to do with one another, arrived at the same musical concept, chord sequence, feel or lyrics at the same time in different places, this is an example of synchronicity.

Saying that the connection is not causal implies the conviction that there is no known type of transmission (energy waves, phenomena of radiation, etc.), and these events defy and even suspend our classical references of space and time. We must realize that causality as an explanatory principle is only one possible category of thought to describe connections between events [40].

Jung calls acausal order those phenomena, physical or psychic, that have an *a priori*, timeless order, which remain always constant, as radioactive decay would be in physics, and the property of natural numbers would be in the psyche. Thus, this acausal order existing in nature would be present in both the psychic domain and in material reality, where it could always be seen as regular, determinable and universally valid [41]. The sporadic events of synchronicity, however, which also manifest an acausal order, are unexpected, irregular and occur unpredictably, and, according to Jung, would be a particular case, a manifestation of the general principle of acausal order when it is inserted in the linear and causal time usually experienced by our conscious psyche [42].

As for the equivalence of meaning, this possibility is based on the existence of an absolute and transcendent knowledge. I choose Jung's own words:

> Absolute knowledge, which is a characteristic of synchronistic phenomena, knowledge not conveyed by the sense, underlies the hypothesis of meaning subsisting in itself, or expresses its existence [43].

This absolute knowledge, existing in the psychoid scope of nature, would be responsible for the structural significance that would manifest equivalently in the psyche and the physical world [44]. The human psyche, in turn, through the reflective consciousness, becomes the receiver of such standards and gives them an elaborate expression, allowing conscience expansion and deepening of the meaning of human life. Despite being isolated, non-reproducible events, and thus non-verifiable, synchronistic events have a huge value of information to the individual and, consequently, to the cosmos as a whole [45].

Naturally, another question arises: Can we, through some underlying law, predict the occurrence of synchronistic events? The answer, to this day, remains negative. The only factor detected within a larger trend of occurrence of synchronistic events was the intense activation or constellation of an archetype [46].

As pointed out by Von Franz:

> Jung therefore describes genuine synchronistic phenomena as 'parapsychological,' marginal phenomena which are only observable when our ego consciousness becomes 'dimmed.' This would mean that the luminosity of a constellated archetype, which shines forth in the 'meaning' of a synchronistic event, increases its energetic charge in proportion to the degree that the light concentration of ego consciousness diminished ... [47].

On such occasion, if an external event follows the same basic pattern, behaving in the same way that the constellated archetypal structure, that physical event will be equivalent to the psychic event in terms of meaning. Thus, we will be able to recognize in the phenomenon, the manifestation of the psychophysical dimension that most of the time we are unable to see. This leads us, mistakenly, to believing in the independence between the physical and psychic worlds. The connection through equivalence of meaning tells us, as in the Chinese thought, that nature as a whole is a psychophysical unit that Jung named *Unus Mundus*, which will be discussed below.

To finish, a story that originally appeared in Life Magazine:

> All fifteen members of a church choir in Beatrice, Nebraska, due at practice at 7:20, were late on the evening of March 1, 1950. The minister and his wife and daughter had one reason (his wife delayed to iron the daughter's dress) one girl waited to finish a geometry problem; one couldn't start her car; two lingered to hear the end of an especially exciting radio program; one mother and daughter were late because the mother had to call the daughter twice to wake her from a nap; and so on. The reasons seemed rather ordinary. But there were ten separate and quite unconnected reasons for the lateness of the fifteen persons. It was rather fortunate that none of the fifteen arrived on time at 7:20, for at 7:25 the church building was destroyed in an explosion. The members of the choir, Life reported, wondered if their delay was "an act of God."

Chapter 7

UNUS MUNDUS

The psychic reality introspectively described by the analytical psychology as the collective unconscious seems to be equivalent to the reality explored by the atomic physics, regarding it from the outside, as a 'material' reality [48].

In both sciences, elements to identify clearly the connection between them are still missing. It is worth noting, however, that neither the material nor the psychological worlds need to be derived from each other. They show, more and more, that they are coordinated and equivalent domains possibly reflecting the mystery of a unit that escapes us. The Latin expression *Unus Mundus*, created by the alchemist Gerard Dorneus, a disciple of Paracelsus who lived in the sixteenth century, was borrowed by Jung to name this unitary reality of the physical and psychological domains.

As stated by Jung:

> The idea of *Unus Mundus* is based undoubtedly on the assumption that the multiplicity of the empirical world rests on an underlying unity and that no two or more distinct worlds exist side-by-side or intertwined with one another. In fact, everything that is divided and different belongs to the same world, which is not the world of the senses... [49].

Thus, the empirical reality would be based on a dimension both physical and psychic and, thus, it would be better characterized by a third possibility, the *Unus Mundus*, fundamentally transcendent in essence and which allows us to perceive only clues.

As explained by Von Franz:

> Jung used the expression *unus mundus* to designate the transcendental unitary reality underlying the dualism of psyche and matter. The idea of such a unity behind all existence is based on an archetypal foundation. The expression *unus mundus* originated in medieval natural philosophy, where it denoted the timeless, preexistent, cosmic plan or antecedent world model, potential in God's mind, according to which he realized actual creation [50].

The assumption of the existence of the *Unus Mundus* would stem from sporadic manifestations of the synchronistic phenomenon, where the psyche-matter duality seems to cease to exist. As similar structures manifest themselves both in the unconscious psyche and in matter, the unity of existence which underlies the duality of psyche and matter becomes more comprehensible to us.

In Von Franz words:

> Although the nonperceptual potential continuum or *unus mundus* appears to exist outside time, certain dynamic manifestations of it break through into our ordinary temporal sphere in the form of synchronistic occurrences [51].

Moreover, the equivalence of meaning in this synchronicity would also point to this psycho-physical unity of absolute knowledge that would transcend, altogether, our conscious awareness [52].

This absolute knowledge of the *Unus Mundus* would be a factor of structural significance, full of meaning and responsible for an acausal order of nature, of which synchronicity would be a particular manifestation. In addition to absolute unity, timelessness, characterized by what contains simultaneity outside of time, would also belong to the *Unus Mundus*.

This eternal, of utter meaning, would enroll sporadically in our linear time and in our causal world in the form of equivalences of psychic and physical significance, and would open in a numinous way a "window of eternity" where we would be allowed to experience the transcendence of the *Unus Mundus*.

In Von Franz words:

> Since the concept of the *Unus Mundus* transcends consciousness, it is represented in mankind's historical Weltanschauungen by symbols, which most frequently consist of a double mandala portraying both the timeless

> and time-bound order of existence. While the timeless order seems to relate to the general concept of acausal orderedness in the physical and psychic realms, the time-bound order refers more to peripheral phenomena, such as synchronistic happenings, that are creative acts in time. The timeless acausal orderedness lies at the base of all transmittable information and cognition processes operating in man, and the time-bound synchronistic phenomena underlie those individually experienced messages of the unconscious which can only be adequately interpreted by the individual. The two systems are incommensurable, and because of this they form a fitting symbol for the ultimate unity of existence as a coincidentia oppositorum. In their mirrored images, as one reality reflects off the other, lies the mystery of their experience ability by the individual [53].

Finally, Von Franz teach us that synchronistic events appear to be linked up with an individual's inner development and in some way dependent on it [54].

She gives us two examples:

> In the course of observing the effects of activated archetypes or patterns of psychological reactions in his patients, Jung again and again encountered cases where rare outer chances tended to coincide meaningfully with archetypal dream images. Such cases seem to occur only when an archetype is activated producing highly charged conscious or unconscious emotions. The following may serve as an illustration: A woman with a very strong power complex, and a 'devouring' attitude toward people, dreamt of seeing three tigers seated threateningly in front of her. Her analyst pointed out the meaning of the dream and through causal arguments tried to make her understand the devouring attitude which she thus displayed. Later in the day the patient and her friend, while strolling along Lake Zurich, noticed a crowd gazing at three tigers in a barn – most unusual for a Swiss barn!
>
> Taken separately, the causal background of the two events seems clear. The power display in the woman probably brought about her tiger dream as a warning illustration of what was constellated in her. The three real tigers was in that barn because a circus was spending the night in town. But the highly improbably coincidence of the inner and outer three tigers in the woman's life seems to have no common cause, and therefore it inevitably stuck her as 'more than mere chance' and somehow as 'meaningful' [55].

The other example, in Von Franz words:

> I once received the following communication (which I have permission to make public) from a young Jewish doctor, an internist stationed in Hawaii. He had intended to spend a day off preparing himself for his nephew's *bar mitzwah.* Since he was completely unacquainted with the ceremony, he began to read some of the sacred texts, but he became bored and reached instead for my book *Individuation in Fairy Tales*. In this book I interpret a Spanish fairy tale in which the hero's task is to catch hold, at the right moment, of a white parrot that guards the source of life. Whoever tries to catch hold of it at the wrong moment is turned into stone, petrified. The hero succeeds in catching the bird, and the white parrot later becomes the protective spirit of the hero and his family. In Junguian terminology, it is a symbol of the center that encompasses the wholeness of the psyche that Jung calls the Self. The negative petrifying aspect of the parrot is related to the danger that threatens all religions – but especially religions of the Book – that their insights will be learned and repeated only in a parrotlike manner. In such a case, religion, instead of having a healing effect, has a petrifying one.
>
> So the spiritual content of religion must be "grasped" in the right way for its positive impact to be reached. This interpretation directly touched the young doctor, since he has just been intending to learn the sacred Jewish texts and formulas by heart in a soulless, parrotlike manner.
>
> At his moment he heard a cawing and chattering outside his window. He looked out, and there in a bush opposite the window sat a fat white parrot! It had a broken golden chain on its foot, having escaped from its owner. It is true that an escaped parrot can survive better in Hawaii than in Europe, but it is still far from normal to find them just flying around! It is scarcely conceivable that reading the fairy tale about the parrot had attracted it causally, and it is likewise inconceivable that the parrot had brought about the reading of the fairy tale. Thus a causal relationship is highly improbable. Therefore, we are led to understand such an event as a meaningful coincidence. That the symbol of the magical bird is a very widely appearing archetypal image is something I cannot document here; I must refer the reader to my book. But that a repressed religious problem - and thus an archetype – was touched here in the young doctor is obvious. This example is very useful in explaining why Jung speaks of equivalence rather than meaning when no experiencing subject is included in an incident. If no observer were present, there would be in this story only two equivalents parrots, one inside and one outside.

> Meaning reveals itself only when an observer is present. It is experienced of a kind of miracle, but the scientists would mostly push it aside as a meaningless chance event [56].

This brings the necessity of dealing with the problem of meaning. Meaning was understood by Jung not as something partial, like the meaning of a sentence but rather as the meaningful coherence of being as a whole. The closest thing to this is the Chinese concept of *Tao.* In the West, the *logos* fire of Heraklitos and then the *nous* as cosmic intelligence, as it is described in the *Corpus Hermeticum*, are the concepts that come closest.

Synchronicity is soul-making, a revelation of the cosmic self or soul as meaning. If we have the inner eyes to see, the self is providing us with both the necessary experience and meaning required for our transformation, our individuation [57].

Chapter 8

COLLECTIVE UNCONSCIOUS AND NUMBER

> "Philosophy is written in this grand book, the universe, which stands continually open to our gaze. But the book cannot be understood unless one first learns to comprehend the language and read the letters in which it is composed. It is written in the language of mathematics, and its characters are triangles, circles, and other geometric figures without which it is humanly impossible to understand a single word of it; without these, one wanders about in a dark labyrinth."
>
> *Dante*

As discussed previously, the transcendent *Unus Mundus* would be the symbolic expression of psyche and matter united, impossible to be grasped by consciousness. Inherent of absolute knowledge, a factor of order and significance, the *Unus Mundus* symbolically brings to the conscience an archetypal element of order and regularity that cannot be characterized in a more fundamental manner than using natural whole numbers. Mathematicians in ancient times already pointed out the divine character of the numbers that we, moderns, fail to grasp. Nevertheless, we can assume this aspect when we observe the structures of order in mandalas, seen as images of the divine. In the words of Von Franz [58], "Jung defined the number simply as the order archetype that became conscious."

Western culture, in this case specifically referring to mathematics, has been developed in order to lose contact with the divine origin of the number, as well as its qualitative aspect [59]. Numbers are treated as creations of the human mind, expressing quantities and relationships between them. But natural numbers are, in essence, unfathomable and inaccessible, each with specific dynamic qualities, a specific rhythm, related to a field and continuum, rather than discontinuous and discrete, objects rooted in the collective unconscious [60].

In the words of Robert Lawlor [61]:

> The irrational functions (which we will consider rather as supra-rational) are a key opening a door to a higher reality of Number. They demonstrate that Number is above all a relationship; and no matter what quantities are applied to the side and to the diameter the relationship will remain invariable, for in essence this functional aspect of Number is neither large nor small, neither infinite nor finite : it is universal.
>
> Thus within the concept of Number there is a definite, finite, particularizing power and also a universal synthesizing power. One may be called the exoteric or external aspect of number, the other the esoteric or inner, functional aspect. Let us look at the first four primary numbers in this spirit.
>
> The number ONE can of course define a quantity; as, for example, one apple. But in its other sense, it perfectly represents the principle of absolute unity, and as such has often been used as the symbol to represent God. As a statement of form it can in one sense represent a point - it has been called the 'pointal' number, the bindu or seed in the Hindu mandala - or in another sense it can represent the perfect circle.
>
> TWO is a quantity, but symbolically it represents, as we have already seen, the principle of Duality, the power of multiplicity. At the same time it has its formal sense in the representation of a line, in that two points define a line.
>
> THREE is a quantity, but as a principle it represents the Trinity, a vital concept which we will meet again later. Its formal sense is that of the triangle, which is formed from three points. With three a qualitative transition is made from the pure, abstract elements of point and line to the tangible, measurable state which is called a surface. In India the triangle was called the Mother, for it is the membrane or birth channel through which all the transcendent powers of unity and its initial division into polarity must pass in order to enter into the manifest realm of surface. The triangle acts as the mother of form. But three is yet only a principle of creation, forming the passage between the transcendent and the manifest realms, whereas FOUR represents at last the 'first born thing', the world of Nature, because it is the product of the procreative process, that is of multiplication: 2 x 2 = 4. As a form, four is the square, and represents materialization.

Is important to know that ancient geometry begins with One, while modern mathematics and geometry begin with Zero. These two symbolic beginnings provide an example of how mathematical concepts are the prototypes for the dynamics of thought, of structuring and of action. Let us first consider zero, which is a relatively recent idea in the history of thought, yet is already so ingrained in us that we can hardly think without it. The origins of this symbol date back to sometime before the eighth century AD, when we have a record of its first written appearance in a mathematical text from India. This is now considered by many to be a dark period in the long, rich spiritual heritage of India, a decline from the previous tradition which upheld a spiritual significance in both the manifested and the un-manifested expressions of God. It was at this time that the concept of zero took on a new tangibility and presence. The result was that it achieved a specific name and symbol in both metaphysics and mathematics. In mathematics it came to be considered just like the other numbers, as a symbol which can be operated upon and calculated with. The Arabs carried this knowledge into the emerging ferment of Western Europe. During these centuries they picked up the concept of zero along with nine other numerical symbols which had developed in India. The less mystical and more practical orientation of the Arab mentality saw in these symbols a practical device for facilitating calculation and recording large numbers, particularly numbers containing an empty column such as 1505. The great eighth-century Arab mathematician, Al-Khwrizmi, (from whose name derived the word algorithm) carried the Indian numerals with zero to the Islamic world. Gradually this 'Arabic' number system was introduced into medieval Europe and began to support radical changes in Western science and thought. The consequences were enormous. The breakdown of the simple, natural logic of the arithmetic structure allowed a complicated mental logic to take its place and invited into mathematics a whole range of numerical and symbolic entities, some of which have no verifiable concept or geometric form behind them. The invention of zero permitted numbers to represent ideas which have no form. This signals a change in the definition of the word 'idea', which in antiquity was synonymous with 'form', and implies geometry. The 'western' rationalistic mentality negated the ancient and revered spiritual concept of Unity, for with the adoption of zero, Unity looses its first position and becomes merely a quantity among other quantities. The advent of zero allows one to consider anything below the quantitative number series as nil or of no account. Hence zero provided a framework in western thinking for the development of atheism or negation of the spiritual [62].

From the point of view of the natural world, zero does not exist; it is a completely mental entity. Yet the impact of this symbol was enormous. Zero continued to be formative in the nineteenth century world view through the idea that there is a separation between the quantitative and the non-quantitative; the extreme degree of this idea was that everything which is non-quantitative is non-existent or zero. Twentieth-century nuclear physics no longer conceives of the atom as a separate attracting and repelling particle, but instead it poses a field or matrix of interconnected, continually transforming energy fields of particles and patterns. Particles indistinguishable from process; matter indistinguishable from events. Likewise in the heavens, what was once thought to be a black, empty void with bodies floating in it is now known to be filled with substance-energy. Between a stellar body and the region surrounding it, there is a field continuum of which the star-body is simply a densification. Today's science shows us a continual fluctuation and alternation between matter and energy, confirming that in the natural world there is no zero [63].

The notion of zero also had its effect on our psychological conceptualizations. Ideas such as the finality, death and separation owe much to the notion of zero. We saw ourselves as separate individuals moving in a space. But these concepts are now also loosing their hold. We know now that we exist in groups, determined by various levels of energetic affinities, repelling, exchanging and absorbing through interconnected, subtle energetic communications [64].

Returning to Analytical Psychology

Since the collective unconscious is simply pure nature, "nature which includes *everything*; thus also the unknown, inclusive of matter" [65], numbers turn out to be the bridge or neutral language physicist Wolfgang Pauli wished to exist to describe the psychic and physical phenomena simultaneously [66].

Jung also said that numbers are "the most primitive expression of the spirit." It is important to say that he was referring to spirit as the dynamism of the unconscious that, together with its aspect of matrix or maternal womb, possessed a dynamic aspect, a facet of movement, and which acted spontaneously, at its sole discretion. It would create and manipulate images, as in our dreams. We tend, with the development of our consciousness, to appropriate these images, believing that we created them, that we are their authors. This happened, as was already explained, for example, with natural whole numbers that, with the evolution of mathematics, are no longer cosmic divine principles, as they were to the Pythagoreans, and became a series of signs supposedly created by the human mind [67].

There is, however, much of the spirit manifesting itself still today in a completely unattended manner since it is not possible, in any way, to exhaust or dominate the objective psyche and its irrepresentable archetypes.

There are even motives that appear in nature itself and which, in my view, are expressions of this universal numerical order.

In Von Franz words:

> "Very often there is the motif, for instance, of artefacts being made which we think only the human psyche can produce, e.g., such carvings in a cave, which are produced there by nature" [68].

Therefore, it is my intention to further analyze one of these motives, called "crop circles," exploring their harmonic geometry and numinous beauty.

Chapter 9

CROP CIRCLES

"Crop circles are a numeric and geometric manifestation of nature expressing quantitatively and qualitatively its eternal order, transcendent and full of meaning. They bear the numinous character of the archetype, impacting us with their harmonic beauty. They do not present themselves without a reason, but within a qualitative time that is pregnant with meaning, when our rationalist and unilateral view of the world begs for mystery."

(Simone Mancini Castilho)

In thc words of thc architect Michael Glickman, extracted from his book "Crop Circles: The Bones of God" [69]:

A photograph taken in 1907 during the Wright brothers' visit to France to demonstrate their flying machine shows a farmer in Pau in the Pyrenees, pitchfork in hand, standing on top of his hay cart gazing skyward as Wilbur, or perhaps it was Orville, passes overhead. We cannot see the farmer's expression. We do not know who he was but it is safe to assume that this was the most shocking experience of his life. It is the most powerful image I know of cognitive dissonance.

A century later, heavier-than-air flight is unexceptional. But image for a moment the response of this bewildered farmer to a view of the world as it *should not be.* Living in southwestern France a hundred years ago, he was probably a religious man and, while he will not have taken this as either an angelic vision or visitation, he might have felt, to invert the phrase, that he had witnessed a *machina ex deus.*

For several years after the Wright brothers' first flight in1903 and as the news of their feat started even more insistently to filter through, the great and the good pronounced definitively upon the simple impossibility of their claims. This was the world as it could *not* be. Despite the fact that hundreds of people have witnessed the flight of a heavier-than-air craft from the public tramway that ran alongside the field, the idea was denounced by authorities as an affront to approved reality, to the consensus view of the world.

And, similarly, it is made clear to us that the crop circles simply *cannot be.*

9.1. General Considerations

The so-called crop circles, or crop formations, are mysterious circular formations arranged in various combinations that appear inexplicably printed on cereal fields by bending the stem of plants in a spiral from the core outward. Crop circles occur in almost all types of grain or cereal crops, including wheat, rye, barley, tobacco, oats and maize, as well as grasslands, rice crops and other vegetation. Since the first report in 1678 there have been over 10,000 circles worldwide, predominantly in Britain. Their appearance has become more frequent since 1989; usually 100-250 formations arise annually in many countries such as England, Canada, USA, Australia, Italy and Japan. Three events were recorded in Brazil, in the years 2008, 2009 and 2012, all in the municipality of Ipuaçu (indigenous community) in the western part of the southern state of Santa Catarina. The flattened appearance of the circles is not produced by cutting the stems of the plants, but by the fact that they are bent at a ninety-degree angle on the root node (the first above ground). Despite the anatomical changes occurring in the plants, they remain alive and continue to grow. The crop circles that appear today have hundreds of feet in diameter and the drawings reveal layers of plants bent and intertwined over one another, producing intricate and geometrically precise patterns. Despite their greater complexity, the formations remain individual and unique.

9.2. The History of the Crop Circle Phenomenon

Crop circles are not a phenomenon of the present. They are mentioned in academic texts of the late seventeenth century and until 1970 some 200 cases were reported, with the vast majority occurring in crop fields located in southwest England.

Until 1980 the shape was essentially a single circle. Then they began to appear as groups of three, four and five circles, suggesting an underlying geometric or numerical order. Until then, the formations were attributed to simple causes as the growth of some kind of fungus, irrigation patterns or strong whirlwinds.

In the late 1980s, the first linear patterns emerged, drawings of insects, and the forms became larger and more beautiful. They continued to be called crop circles although sometimes the terms agroglyphs or pictograms also began to be used.

At the beginning of the 1990s, a woodcut pamphlet carved in 1678 was discovered showing the "Mowing-Devil", which illustrated a popular story about a pattern emerged mysteriously in the fields overnight. This woodcut generated research efforts and a search for historical records of such formations.

After 1990, the complexity of designs increased exponentially, with the emergence of images similar to computer fractals, and elements expressing the four-dimensional models of quantum physics. At that time, more precisely in mid-1991, covered by extensive publicity, a hypothesis of scam or hoax was raised as the cause of all crop circles. Two men in their 60s, Doug Bower and Dave Chorley, presented themselves as the authors of the drawings in the fields, stating that they had done the first circle in 1978, a sort of joke, and declaring themselves responsible for the creation of more than 250 circles since that year.

Despite the inconsistency of their reports and subsequent statements that no longer confirmed authorship, the question of crop circles was "settled" for the media and most people. But even if their claims were to be believed, they were making a maximum of 30 formations a year, leaving the great majority unaccounted for.

That raised lots of questions and in the context of many doubts like: who were all the hoaxers, apart from Doug and Dave? were they in the military or intelligence service or were they members of occult groups?, a group of people, including Rupert Sheldrake, John and Richard Adams decided to organize a crop circle making competition to find out exactly what people could and could not do. The competition was co-sponsored by The Guardian and The Cerealogist, the leading crop circle journal in Britain, edited by John. A German magazine, PM, funded the contest and offered a first prize of £3,000.

Sheldrake was the chairman of the organizing committee and described the competition as follows [70]:

"The competition was announced in May 1992 and teams interested in taking part had to submit an application in advance. A panel of crop circle experts advised us on the design of a standard formation that all the teams were required to make. This contained supposedly difficult-to-hoax features like rings, and the instructions also specified the directions in which the crops were to be flattened. This design and general guidance notes for competitors were sent to all of the participants three weeks before the actual contest, so they had time to plan their strategy. They were also sent a copy of the marking scheme by which the judges would allocate points to the various formations. The location of the contest was kept secret, and the participating teams were informed of it only 48 hours in advance.

The contest itself took place on the night of Saturday July 11, 1992, in a wheat field on the estate of Sir Francis Dashwood in West Wycombe, Buckinghamshire. On Saturday afternoon, we marked out equal-sized plots that were allocated at random to the 12 competing teams. The competitors had to arrive at the tent by 9pm, when they were given the final rules and instructions. They were free to work in their plots in the hours of darkness, between 10pm and 3am, when they had to leave the field. During this period, members of the organizing group were patrolling the area to make sure there was no sabotage of other people's plots by contestants or by anyone else, and after all the teams had left the field, the area was guarded by Sir Francis Dashwood's gamekeepers.

We arranged for a helicopter to fly over the field soon after dawn to take aerial photographs of the site. The five judges, led by Jürgen Kroenig, were in the field soon afterwards. When we were assembling the team of judges, we invited the Wessex Skeptics, who claimed that all crop circles were hoaxes, to nominate a judge, but they refused. They thought that our set design was too difficult to execute in the dark and claimed we were deliberately setting up an unfair contest. We also invited Doug and Dave to take part, pointing out that with all their practice, they should have found it easy to win the £3,000 prize.

They also declined.

The rules had specified in advance that lights could be used, but would result in a loss of points. Only one team worked with light.

The night was cloudy, with intermittent rain and only occasional moonlight. The next morning, after the aerial photographs had been taken, the judges were the only people allowed to enter the field. By midday they had completed their judging process, taking into account both their observations on the ground and the aerial photographs. The winning team consisted of three young engineers from the Westland Helicopter Company. However, before being presented with a cheque for £3,000 by the Earl of Haddington, they had to make the formation again in broad daylight so that everyone could see how it was done.

Programmes about the competition on BBC radio had revealed the location that morning, and the circle-making demonstration was open to the public for a modest entry fee of £1. Several hundred people came to watch, along with TV crews, journalists and photographers.

The demonstration by the winning team was fascinating. The young engineers employed very simple apparatus. For flattening the crop, they used a roller consisting of a piece of PVC piping with a rope through it, pushing it with their feet. In order to get into the crop without leaving footprints, they used two lightweight aluminium stepladders with a plank between them, acting as a bridge. For marking out a ring, they did not put a post in the centre, but rather used a telescopic device made out of plastic pipes of different diameter projecting from the top of an aluminium step ladder. A string was attached to the end of it in such a way that by holding the string and walking in a circle around this central position a perfect ring of flattened plants could be marked out without leaving any trace on the ground in the middle.

The experiment was conclusive. Humans could indeed make all the features of state-of-the-art crop formations at that time. Eleven of the twelve teams made more or less impressive formations that followed the set design. The contest was reported in most of the British newspapers and on radio and TV.

Some people jumped to the conclusion that all crop circles must be hoaxes. The sceptics were pleased with the outcome, although they had been so skeptical about the competition itself and had refused to cooperate in its organization.

After the competition I took part in many interviews, in which I pointed out that although the results showed that it was possible to hoax crop circles, it did not prove that all were hoaxes. The fact that it is possible to forge a £50 note does not prove that all £50 notes are forgeries.

The same night that the contest took place, three other crop formations appeared in fields in southern England. One of them was five miles from West Wycombe.

Since the competition, formations have continued to appear. Although the numbers have decreased, the designs have become ever

> more sophisticated, many of them embodying remarkable geometric constructions. Every year the Wiltshire Crop Circle Study Group produces a magnificent calendar showing the best crop circles from the previous year, and I have one on my study wall, reminding me daily of these extraordinary formations, with their spectacular designs and dazzling displays of creativity.
>
> Almost everyone now agrees that most crop circles are human made. But some enthusiasts still believe that a minority are created by non-human agencies. Surprisingly, there was an article about them in the scientific journal Nature on June 10, 2010, called ‘The Crop Circles Evolves.’ The summary reads: “A growing underground art movement combines mathematics, technology, stalks and whimsy.” But even this scientifically acceptable account has not managed to expel all mystery. The details of the bent stems suggest to the author of the Nature article that “some patterns may have been sculpted using microwave generators, such as masers or magnetrons from microwave ovens.”
>
> John enjoyed the continuing evolution of crop circles, their increasing geometrical sophistication, and the way they continue to defy simplistic explanations.”

In 1994, other group of three individuals who originally called themselves “Team Satan”, and later “The Circlemakers”, decided to claim authorship of some of the most elaborate crop circles in England. Attracting much media attention, earning lucrative commissions and sponsorship of private companies, the group has been invited to create geometric patterns and logos on crop fields, possibly aiming to acquire notoriety and trivializing the phenomenon. The fact that many of the drawings in fields can be made by men should not prevent us from seeing that formations produced like that fail to reproduce any of the atypical biophysical and electromagnetic characteristics proven to belong to the authentic phenomenon. The testimony of dozens of people who claim to have witnessed the appearance of crop circles in seconds or minutes, the fact that footprints or marks on the ground were never found and that sometimes the appearance of the drawings took place in broad daylight are information not considered if we admit exclusive creation by human hands. Crop Circle Challenges funded by the media have already been launched. Even with results that fail in terms of geometric accuracy and which are achieved with the destruction of the plants and requiring hours to days, such challenges seem to achieve the purpose of responding to the fearful immediacy that seeks prompt and definitive answers. A pearl made by human hands does not prevent there being another one found inside an oyster at the bottom of the sea.

Humanity seems largely to dodge a more careful look at what manifests in nature, insisting not to see the fact that the plants are not cut or damaged, but bend to ninety degrees, close to the ground, and remain in this position while maturing. The option is blindness rather than admitting uncertainties and confessing that the truths of science must be reassessed. In fact, men can create crop circles with boards and ropes, but we cannot ignore those offered to us as gifts, the deservedly called authentic circles.

9.3. THE INCREDIBLE CROP CIRCLES

Again, I will borrow a fable from Michael Glickman's book [71].
In his words:

> Imagine.
>
> Our huts are set around a small clearing in the rainforest on the bank of a boundlessly wide river. We have never crossed the water. It is the edge of the world and we live in fear of what might be on the other side. We are a small group and we know no others. The elders have spoken of a people far up the river. They tell us that in ancient times they used to raid us but none of us remembers this. We never see them. They do not exist for us.
>
> We heard a strange roaring and went to the water's edge to see.
>
> There seemed to be, far away, a canoe. But it was too big for a canoe. It roared like a creature and it farted smoke. There were figures standing on the creature's back and we thought they were men. But as the beast swam closer we saw that they were big and pale and wore strange garments.
>
> We ran into the forest and watched, hidden in the trees. The creature swam to our clearing and the pale men got off its back and walked into our village. They called out to us, but, of course, we hid ourselves.
>
> "Let us kill them", said the young men.
> "If you kill them more will come", said the elders.
> "If we hide and keep silent they will go", said the women.

They waited for a day in our clearing. They held smoking sticks in their mouths. When they saw that we would not come, they left trinkets on the ground and went back down the great river.

We let time pass and then slowly moved back to gaze at their strange offerings. There were gaudy beads and shiny things. We were all afraid to touch them and stood in a circle gazing. The shaman took bones from his pouch and cautiously used them to tap on each of the objects, chanting all the while. Then the head man stepped and picked up one of the shiny things. We learned that if you looked into it, you saw a face. It was disturbing. And more so when we came to understand that it was our own face!

For some days the gifts lay undisturbed in our clearing. We would stand there and stare at them and, little by little, we found the courage to pick them up. The head man chose two or three items to take to his hut. Soon the marvelous become the familiar. The little pile of offerings was dispersed and the clearing was empty.

We never fully understood what had taken place. But now in some way we are all different.

As surprising as the phenomenon of crop circles is how difficult it is for the world to perceive it.

The huge and spectacular designs with sophisticated geometry, mathematics of fractals, numbers and shapes of the universe, appearing stamped in the fields of grain, reveal to ourselves our so narrow vision. The subtle intertwining of plants, forming layers with thickness, colors and reflections that embellish the fields in stunning aerial images, definitely proves our severe blindness to the unconceivable. Full of arrogant ignorance, we choose to naturally assume their obviousness, or radically reject them. And being literally amazing, we conveniently call them "fake", in a clear display of our difficulty in admitting that an unknown phenomenon is authentic as, in my view, crop circles are. Hastily, we judge before anyone else can call us foolish or naive.

Here, the words of philosopher Soren Kierkegaard fit well: "There are two ways to be fooled, one is to believe what is false; the other is to refuse to believe what is true."

History has many examples of figures who were convicted of raising different and unprecedented points of view. Researchers and their findings about crop circles are not different. Many newspaper and television reports do not disclose historical or scientific evidence even when they exist. We know the psychological distress caused by what is new and how ready we are to deny it. We must bear in mind that our internal taboos can make us ignore new points of view, condemning us to reverberate our old and accommodated concepts.

Our ability to experience admiration is numb, lulled by our intellectual arrogance. The harmonic beauty of the designs, however, resists our limitation and revives every year, giving us another chance to regain our senses.

9.4. Authentic Crop Circles: Physical, Cellular and Germinal Changes

The physical changes in plants in crop circles have been documented by Michigan biophysicist W.C. Levengood [72] after evaluating hundreds of plant samples - both downed and standing - taken from within the perimeter of each formation and compared to hundreds of control plants taken at various distances away from them, in many directions. More than 250 individual formations in crops located in different countries over a period of ten years were thoroughly examined. Although many of the formations studied were relatively "simple" in overall design and/or relatively "small" in overall size, a number of more "complex" or larger formations - those whose overall design included intricate geometric shapes with multiple design elements of varying sizes - were also represented.

Many of the formations were from European countries and, in these cases, the plants were allowed to dry in the open for 4 to 6 weeks before being sent to the United States. Some of the formations coming from the US and Canada were sent while still green. See below the physical changes found in plants from crop circles, statistically significant using a confidence interval of 95%:

- Laterally and longitudinally enlarged plant stem nodes.
  The nodes, fibrous protuberances, are the hardest portions of the stem, allowing a plant to support its weight and maintain its upright posture.
- Marked bending of the plant stem nodes
- Marked bending at the base of the plant stem
- Expulsion cavities (holes blown from the inside out at one or several of the plant stem nodes).

Cellular changes:

- On the microscopic level, abnormal enlargement of cell wall pits in bract tissue (a thin membrane which surrounds the seed- head and through which nutrients pass to the developing seed)

Germinal changes

- Absence of seeds
- Stunted, malformed seeds
- Up- or down-regulation in germination patterns

In Silva words [73]:

> Levengood concluded that 'the energy mechanism producing quantitative alterations in the plant stem nodes falls within the framework of a straight-forward and widely applied principal of physics [Beer's Law] dealing with the absorption of electromagnetic energy by matter', strongly suggesting that an energy source 'originating in the microwave region' had boiled the water inside the plants' nodes, effectively transforming it into steam.

Still according to Silva [74]:

> Michael Chorost reached similar conclusions in his published report for Project Argus: the phenomenon induces radiation anomalies, heats plants rapidly and briefly through a rapid pulse of unknown energy, sometimes scorching them; it swells their cell wall pits and interacts with the development of the seeds, and leaves radioactive traces in the soil.

9.5. Other Relevant Aspects

In studies of the phenomenon, biochemical changes in the affected plants were also reported. These effects appear to derive from a process of rapid extreme heating.

Changes in electromagnetic radiation are also observed in areas displaying formations, with frequent reports of malfunction of electrical equipment, new batteries discharged, and magnetic compasses that lose their direction.

Magnetite particles of iron were found on the plants and in the soil under the formations. As in the case of changes in the length of the nodules, the presence of these particles follows the same linear distribution from the center to the periphery of the formation [75].

In addition, the formations occur only in areas where a source of groundwater is near the surface. The appearance of crop circles near sacred sites is also a finding.

A curious fact is the report of bright spherical objects called "balls of light" seen on and around the formations, during the day or evening, immediately before the formations appear. Such images have been witnessed many times and even captured on video. These small sources of electromagnetic radiation ("balls of light") are directly involved in the creation of crop circles. Nevertheless, their origin and true character remain unknown [76].

Visitors of crop circles report varied physical and emotional sensations, such as headaches, disorientation, nausea, well-being or fear.

According to Silva [77], the crop circles were predicted to produce certain physical effects on people, and hundreds of people have experienced healings, unusual clicking sounds, and unusual metabolic effects when visiting the sites. Animals also behave differently while in the formations or avoid approaching them.

Last, and in my opinion of utmost importance, are the reports of simultaneity between psychological processes and the appearance of crop circles. Several researchers have experienced such interaction and reported the appearance of a crop formation that was similar to a prior psychic image they had had.

Other occasions of crop formation appearance have been reported by researchers who had dreamed or thought about a particular image the previous night, or had discussed certain design as they flew to a particular location, finding it stamped into the field minutes later.

Freddy Silva, a researcher of sacred sites and ancient systems of knowledge, reports such occurrences [78]:

> In two separate incidences both the crop circle researcher Andrew Colin and pilot Busty Taylor predicted actual events; Colin had dreamt of a Celtic cross pattern, never before seen. The following morning a local farmer telephoned him to report a Celtic cross crop circle had materialized in a field beside Colin's home. In Busty's case, he had been doing aerial reconnaissance with a colleague when he remarked, "all we

> need right now is to see all the designs that have appeared so far rolled up into a Celtic cross." The next day he flew over the exact same field where 24 hours earlier he had uttered this remark and below lay a perfect Celtic cross.
>
> They were not the only ones to have predicted designs before they appeared in the fields. Noted author Barbara Hand Clow vectored an unusual design as she began to work on a new book related to consciousness. The next day, the same design appeared beneath Barbury Castle- the most complex design to date, one of the most important esoteric, philosophic and cabalistic symbols.

Likewise, there are testimonies of the appearance of crop circles by groups of people while chanting mantras or seeking meditative and psychic expansion states.

In Freddy Silva words [79]:

> In 1992 a scientific experiment was conducted to see if some type of subconscious communication was indeed going on between certain individuals and the Circlemakers. In a closed experiment by Steven Greer/ C.S.E.T.I., four researchers, and psychic Maria Ward, the team agreed to project a design. Sitting at night at the summit of Woodborough Hill in Wiltshire (which coincidentally stands on a geodetic line connected to Avebury stone circle), the group spontaneously agreed to project a pattern consisting of three circles arranged in triangular fashion. Maria suggested adding straight lines to the center of each circle at the last minute. Upon agreement the thought sequence began. The following morning a report came in of the projected image having materialized some miles away as a crop circle, below Oliver's Castle (also on a major geodetic line). The match was identical. Mathematician Gerald Hawkins, who had already proved that certain crop circles contained geometrical theorems based on Euclid, proved this crop circle contained a diatonic ratio, a mathematical relationship never found in hoaxed patterns. But the story doesn't end there. Across the Atlantic, a gentleman in New York State had previously contacted CPRI about a dream on June 28 that a crop circle consisting of three circles connected by three paths within an equilateral format would appear in England on the 23rd July. This was indeed the same pattern that was discovered on the morning of the 24th during the CSETI/CPRI operation.

In these situations, we cannot help but notice the synchronistic phenomena discussed above. Based on this understanding, we must remember that psychic images are not derived from matter in the same way that crop formations do not derive from the psyche. They are, however, manifestations of a single, transcendent instance, called by Jung *Unus Mundus* and known to the Chinese as Tao.

Richard Wilhelm, missionary and sinologist, who wrote the most complete translation of the I Ching from Chinese to German, told C.G. Jung a story he experienced in the early twentieth century. It is the story of the rainmaker of Kiao Chan, a wonderful psychological parable about synchronicity and its affinity with the Tao.

There was a great drought. For months, not a drop of rain had fallen and the situation became catastrophic. The Catholics made processions, the Protestants prayed and the Chinese burned incense sticks and fired cannons to scare away the demons of drought; all without any results. At last, the Chinese said: "We will seek the rainmaker."

And a skinny old man came from another province. The only thing he asked for was a quiet little house somewhere, where he locked himself for three days. On the fourth day, clouds formed and there was a big snowstorm, at a time of the year when snow was unexpected, having fallen in unusual quantity. The excitement in the city on account of comments about the wonderful rainmaker was so great that Richard Wilhelm went to the old man to ask how he had done it. He asked in a typically European way: "They call you the rain-maker, could you tell me what you did to snow?" And the little Chinese replied: "I did not do it, I am not responsible." "But what did you do during those three days?" "Oh, this I can explain. I come from another country where things are in order. Here, things are in disarray; they are not as they should be, in accordance with the arrangement of the sky. Therefore, the whole country is out of the Tao, and I am not inserted in the natural order of these things, either, because I find myself in a country in disarray. So I had to wait three days to be back in Tao; then naturally the rain came."

Jung comments on this story in his Vision Seminars [80]:

> "This is how eastern people think – not causal but synchronistic. The rainmaker thought that having come to a place where there wasn't any rain, something was wrong with the place, but also with him because he was at that place. So he did what we should do when we feel wrong: try to put ourselves in order. This may take three days - 5 or 6 days - it does not matter; but when he felt connected, the rain fell. Is it not that he made

the rain fall, he simply waited until it fell. But the interesting thing is how he truly worked in consonance with nature."

Jung also tells us how it is to be in harmony with the Tao:

> The rainmaker, realizing the wrong situation, turned his attention inward and sought to find the Tao within himself, which ended up happening externally as well. When he was focused, rain fell. It is a miracle only to those who think causally. But if we think psychologically, maintaining a correct attitude makes the right thing happen too. The point is not that someone does the right thing; it is simply the right thing, and the person feels that it had to happen like that [81].

9.6. Understanding and Visions

The first description of the phenomenon comes from an English pamphlet dated 1678, which is a description very similar to the one farmers make today when they come across one of these formations in their crop fields. What seems to have changed over time is the increased quantity and complexity of the patterns and, undoubtedly, the explanation given for the phenomenon.

In this sense, the first description, as well as those adopted in the following centuries, link the appearance of crop circles to demonic powers, witchcraft and evil spirits. Closer to modern times, causal explanations for the phenomenon became those of natural order, such as climate and theories related to the winds. Currently, the variety of explanations is even broader, including mathematical and musical theories, and other ideas involving the Earth's/Gaya's environmental energy, alien or alien-technological, military-governmental, spiritual, geometric and biblical explanations. They all represent attempts to account for the phenomenon [82].

As a matter of fact, what scares is not the attempt to understand. What is absolutely shocking is the fact that the phenomenon remains invisible to most of the world; a world currently overflowing information, connected at unprecedented technological speed. There is a blind resistance regarding that which does not fit in our rational understanding, and the mystery is comfortably replaced by the desperate need of modern men to believe that the phenomenon is in fact a hoax. In the words of Jung, "...the 'civilized' man reacts to new ideas constructing psychological barriers that protect him against the shock caused by innovation [83].

"We need to see the beauty, sigh with fascination, and silence in amazement once again.

Fly amidst our doubts; overfly the art; listen to what it says, this sea of images. Almost a mirage...

What do the crop circles ask us?

Perhaps a gentle look, from the heart. Our essence and our certainty of being part of this mystery, to which we owe our sincere reverence."

(Simone Mancini Castilho)

Chapter 10

CROP CIRCLE AND SELF

> "Little by little through the ages, as man's psychic life has become more and more symbolic. This is naturally a great step in advance, but it carries also a danger, for if the ritual loses entirely its connection with the instinctual sources from which it arose, it loses its power to renew or redeem, for it has become merely an abstraction and has lost its connection with the primal sources of life. In these days, when we have grown so far away from our humble beginnings, we need to contact again these life-giving symbols."
>
> *Esther M. Harding*

In order to understand something about the crop circles we have to look at their geometry and number. The essence of the crop circles is their shape, measure, pattern and proportion. It is their very structure and language [84].

Until 1980, the form of a crop circle was generally that of a circle. Since 1990 the circles and rings began to include other non-circular shapes, such as line pictograms, forms of insects, images similar to DNA helices, and fractal designs. In the following decade, images even larger, with great precision and harmony, appeared carrying meanings of religions from all over the world, as well as mythological traditions. Today, we recognize in the images the presence of rules considered sacred geometry, found in the harmony of shapes seen in nature itself.

The perception of mandalic images in the crop circles is quite immediate. Mandala is the Sanskrit word for magical or sacred circle.

The circle is the oldest known graphic representation, and since the earliest attempts at expression, it emerges trying to give shape to our astonishment in face of Creation. Related to the sphere, the most perfect solid geometric form of our universe, the circle is also a recognized symbol of wholeness and indivisibility. In Jung's words [85]:

> Mandala means 'circle'. There are many variations in type ... Its basic motif is the premonition of a center of personality, a kind of central point within the psyche, to which everything is related, by which everything is organized, and which is itself a source of energy. The energy of the central point is manifested in almost all irresistible compulsions and urges to become what it is, just as every organism is driven to assume the form that is characteristic of its nature, no matter the circumstances.

A remarkable fact is that the psychic images of wholeness, spontaneously produced by the unconscious or the symbols of the Self are mandalic images. They are important from a functional point of view because they not only express order, but also create it. That's why they often appear in times of psychic disorientation, to compensate for a chaotic state or produce numinous experiences. Such ordering schemes are spontaneous products of the unconscious that can be imitated by conscience, but do not originate from it. There is here an analogy with the great concern regarding the authenticity of crop circles. In my view, the formations can be created by human consciousness and will, even though these are not their true origin.

The Self cannot be fully expressed in psychological terms because it is unthinkable in itself. It is the archetype related to our ability to grow as individuals, and also to our call towards the divine. Due to its ordering effect, Jung [86] compares the Self to a circle whose center is everywhere and whose circumference is nowhere. It represents our connection with a higher spiritual order and wholeness. The images of crop circles are symbolic manifestations of the Self that can reorder us within this wholeness. They express a sacred order that can help us reconnect with the divine in matter, thus allowing us a fuller psychic state.

In 1959, Jung wrote, under a psychological perspective, about things seen in the sky, which he called modern myth, and that began to appear frequently in the years that followed the World War II. Circular or lenticular images were seen in the sky worldwide, particularly in the United States. Jung understood that the appearance of these images was essentially based on a foundation of emotional tension present everywhere, that is, a general psychological situation whose origin would be related to a context of collective danger or calamity. Those would be unconscious projections that, since they could not be integrated into consciousness, would produce visions, beliefs and unexpected and inexplicable illusions. Applying to spherical or round objects the same reasoning used to explain dreams, Jung understood them as an analogy to the mandala, a symbol of wholeness, present in all times and places. The mandala, as already mentioned, describes the psychic totality composed of conscious and unconscious; it is a symbol of individuation and archetypal character, representing the image of God and being also a symbol of the soul. As said above, this archetype also presents an ordering role aimed at greater possible unity and wholeness of the personality, having a key position in the unification of seemingly incompatible opposites, proper of the mental dissociation of that time [87]. After little more than fifty years, the understanding of Jung remains valid and can be applied, in my view, to the phenomenon of crop circles.

The mental dissociation "of that time" remains with us, equally or more intensely than before, distancing, without measure, our plans from our desires. The unidentified objects of the past, projected and seen in the sky, seem to be today in the fields of grain, known by the name of crop circles.

The intensity with which the development of our consciousness made us unilateral individuals, brutally disconnected from our divine essence and the mysteries around us is undeniable. Our deterministic and rationalistic science has become our truth. We cannot bear the irrational, and feel like we are the masters of our own destiny. We made of our ego, our totality. We have gained power and lost direction. Crop circle mandalas do not fit into our narrow and frightened views, and are thus judged before even being noticed or admired. Their symbolic, geometric and numerically divine content is undervalued against the obsessive questioning about their origin and authenticity.

The fact that we know that groups of people are devoted to creating them does not make the crop circles less surprising or significant. On the contrary; some questions must be promptly asked, such as: what is it that motivates the creation, often anonymous, of such works of art? Despite the most recent and obvious business purpose of some of the designs, why do mandalic images, ancient and sacred symbols, mathematical proportions and accurate geometries keep appearing? Would it be that we are being challenged to understand our psychic projections of wholeness in numeric patterns appearing in grain fields? Or challenged to recognize the psychoid objectivity of the archetype in the geometry of a snowflake, in a spiral of growth or in our own psyche, a mirror of the universe?

As pointed out by Jung [88], an unknown phenomenon being the external cause of a myth or not does not change anything, since it is not the cause what explains it. Thus, regardless of a causal explanation for crop circles, we cannot turn our eyes away from the myth and the symbolic content of wholeness that they present us. Nor can we fail to realize their persistent and increasingly frequent occurrence in these times of faith in the material world and in the power of human beings. Such unilateral and threatening rationalism, while not made aware or integrated, keeps us in a state of collective emotional tension in which unconscious contents can only be indirectly expressed through symbolical beliefs or expressions that we cannot justify [89]. Therefore, the mandalic and sacred contents of the crop circles would aim at compensating, synchronistically, whatever is lacking to the dissociated collective consciousness of our times. These aspects would be undoubtedly a chance to perceive the great; to reconnect us with both the whole and the Self. If the crop circles can make us humbly bow to the humus of the earth, we will certainly have the chance to regain our wholeness.

"We come nearest to the great when we are great in humility."

(Rabindranath Tagore)

Chapter 11

CROP CIRCLE, NUMBER AND GEOMETRY

KEPLER, JOHANNES: "The laws of nature are but the mathematical thoughts of God."

LEIBNIZ, GOTTFRIED WILHELM: "Mathematics is the honor of the human spirit."

PYTHAGORAS: "The numbers govern the world."

PLATO: "God is the great geometer."

At the entrance of Plato's Academy there was an inscription that read: "Only those familiar with Geometry can be accepted here."

The number and geometry of the crop circles seems to express timeless, universal constants and elements found throughout the natural world, in everything from the tiniest flower to vast star-filled galaxies, and even in the design of our own human bodies [90].

Jung defined the number simply as the archetype of order that has become conscious. Our fascination with numbers is caused by the very nature of the archetype, carrier of the numinous and the sacred. Numbers are special and, allow me to say, mysterious. Jung tells us [91]:

> If, as stated in any manual of mathematics, a group of objects is deprived of all of its features, in the end there will still be its number, which seems to indicate that the number has an irreducible character... The series of integer numbers is unexpectedly more than a juxtaposition of identical units: it contains all mathematics and what else can be discovered in this field... Numbers have always been used to characterize any numinous object, and all numbers from one to nine are "sacred"... The most basic quality of an object is to be one or multiple. Numbers

> help us, before and above all, to bring order to the chaos of appearances. It is the appropriate instrument to create order or to apprehend a certain already present regularity, although as yet unknown; that is, a certain order of things. It is the most primitive ordering element of the human spirit, so that the numbers from one to four are the most frequent and the most widespread since the early ordering schemes are predominantly triads and tetrads.

Von Franz [92] develops this idea and says that the number is not just a construct of consciousness but above all a constituent element of nature. When we takes into account the individual characteristics of the natural numbers, we can see that they produce the same ordering effect on both the physical and psychic worlds, composing the most basic constants of nature expressing the unitary psychophysical reality.

Some images formed in crop circles refer numbers.

Among the many examples of crop circles in which we can appreciate the numeric element, two formations are quite illustrative. In one of them we can visualize the Pi and in another, the Fibonacci sequence and the golden ratio.

The Pi number

The circle brings inscribed in itself, in the relationship between its circumference and diameter, the transcendence of Pi, an irrational number, which cannot be written as a finite number or with repeating decimals. Pi is quoted in an Egyptian papyrus dated 1650 and in the Holy Bible.

The passage is in the first book of Kings 7:23:

> "And he made the sea, molten, ten cubits from brim to brim, round all about; and its height was five cubits; and a line of thirty cubits encompassed it round about," suggesting that the builders of the house of Solomon used the value 3 for the ratio between diameter and length of the circumference.
>
> Pi, a transcendental number, with no definite value; it is 3 followed by an infinite sequence of digits without a repeating pattern, with no apparent order or rhythm. The beauty of this is that we have a representation of a disordered infinite contained and described by means of a relationship, within the area enclosed by any circle drawn or imagined. In 1999, with the help of computers, Pi was calculated to an accuracy of over two hundred and six billion decimal places, confirming

the absence of any repetitive pattern, a mathematical demonstration of the chaos that underlies order.

Pi is transcendental; indeterminate and infinite in nature. Between two natural numbers lies an infinite of irrational numbers, which never fill the gap between the natural, and where the entry of the divine becomes possible. To transcend means to go beyond the limits, go to infinity. The transcendent function postulated by Jung, and considered a central point of his theory, discusses the relationship between conscious and unconscious, finite and infinite, human and divine.

Figure 1. Barbury Castle Wilts, June 1st, 2008, Barley.

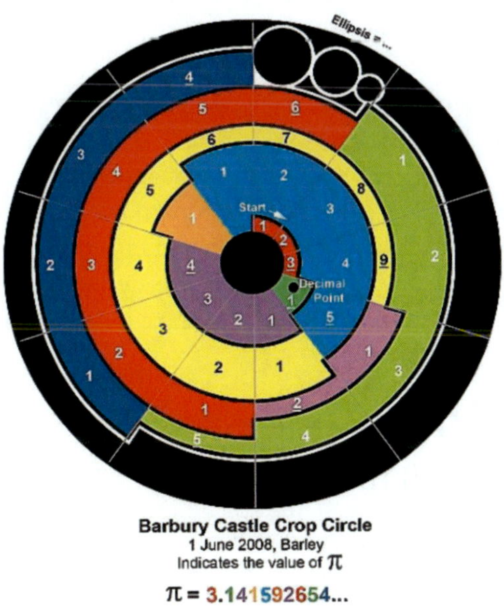

Figure 2. Barbury Castle Crop Circle (Pi Number Diagram).

The Fibonacci Sequence and the Golden Ratio (or Golden Number)

Leonardo Pisano, or Leonardo of Pisa (1170-1250) – also known as Fibonacci - was an Italian mathematician, known as the first great European mathematician after Hellenistic decadence. He is considered by some as the most talented mathematician of the Middle Ages. He became known for his discovery of the Fibonacci sequence and its role in the introduction of Arabic numerals in Europe.

The Fibonacci Sequence

Whereas the first two terms of the sequence are 0 and 1, the other terms are found by adding their two predecessor numbers. Thus:

0, 1, 1, 2, 3, 5, 8, 13, 21, 34, 55, 89, 144, 233, 377, 610, 987, 1597, 2584, 4181, 6765, 10946, 17711, 28657, 46368, 75025, 121393, ...

The golden ratio or golden number

This is an irrational algebraic real constant known by the Greek letter (Phi) and rounded to three decimal places, 1.618. It is a number that has long been used in art and appears in several forms in nature.

It is also called: golden ratio, divine proportion, proportion in extreme ratio, division in extreme ratio. The golden number is approximated by dividing the nth term of the Fibonacci sequence by the previous term. This division converges to the golden number as we take an increasing n.

Construction of a Golden Spiral Using the Fibonacci Sequence

Golden Rectangle

If we draw a rectangle whose ratio of the lengths of the longer and shorter sides is equal to the number of gold we get a golden rectangle.

Golden Spiral or Fibonacci's Spiral

A golden rectangle has an interesting property: if we divide it into a square and a rectangle, the new rectangle is also golden. When this process is repeated infinitely and the corners of the resulting squares are brought together, a spiral is produced, which is then called a golden spiral. This spiral is a curve described in human and animal fetuses.

The proportion that increases the size of the spiral of a snail (not all snails) or the spiral of a Nautilus, which is a type of mollusk found in the sea, is the same as the Fibonacci spiral. The number of petals of flowers in the daisy family is always a Fibonacci number, and the ratio between the number of male and female bees in a hive is also based on Fibonacci. The distribution pattern of sunflower seeds follows two different logarithmic spirals based on the golden ratio, as well as the designs on custard apples and pineapples.

To the extent that plants grow and produce new leaves, the relative position of these leaves will also be based on Fibonacci. The golden ratio was and continues to be intentionally adopted in architecture, especially in sacred architecture, inspiring everything from the size of bricks to global proportions. When we close our hands in a snail shape to grab something, we form the Fibonacci spiral. These are just some examples of the presence of the Fibonacci spiral in nature [93].

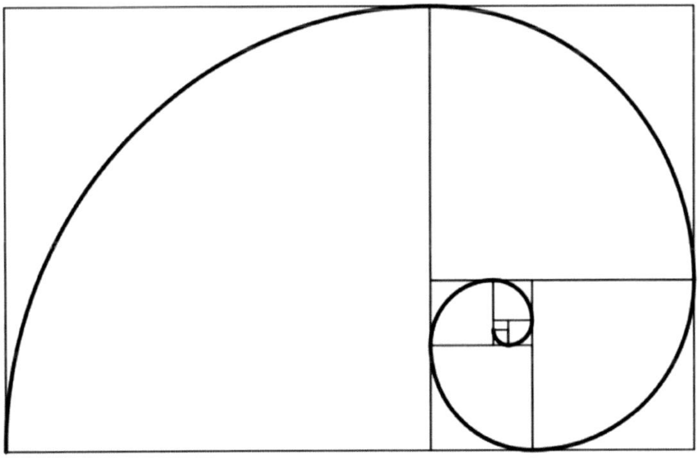

Figure 3. Golden Spiral.

The image of the Fibonacci spiral appeared in wheat fields in Wiltshire, Stonehenge, UK, on July 7th, 1996.

Figure 4. Wiltshire, Stonehenge, July 7th, 1996.

Gerald S. Hawkins, from the department of astronomy at Boston University, a scholar of Neolithic Stonehenge formation in the 1960s, became involved with the analysis of the intriguing patterns revealed in crop circles, devoting his time to an extensive review of all English formations in the period from 1978 to 1988.

He noticed that certain images in crop circles made reference to geometric theorems, expressing specific numerical relationships among the areas of various circles, triangles and other shapes that were part of the composition of a certain pattern. In one, for example, we start from an equilateral triangle that accommodates tightly between an inner circle and an external one. Then, it appears that the area of the outer circle is exactly four times larger than the area of the inner circle.

Hawkins proved four more theorems derived from relationships between areas represented in crop circles using the principles of Euclidean geometry. Also, he discovered a fifth theorem, more general, of which the four previous derived and whose confirmation was launched as a challenge to the readers of magazines *Science News* and *The Mathematics Teacher*. This fifth theorem, that no one could prove, involved a triangle and various concentric circles adjacent to the sides and vertices of the same. It ended up appearing in the geometrical pattern of a crop circle a few years later, in the summer of 1996.

Even more surprising is the fact that all the geometric patterns of the crop circles studied showed exact numerical relationships involving a diatonic ratio, which is simply a ratio of integer numbers determining the scale of musical notes, revealing a connection between geometric harmony, mathematical theorems and music [94].

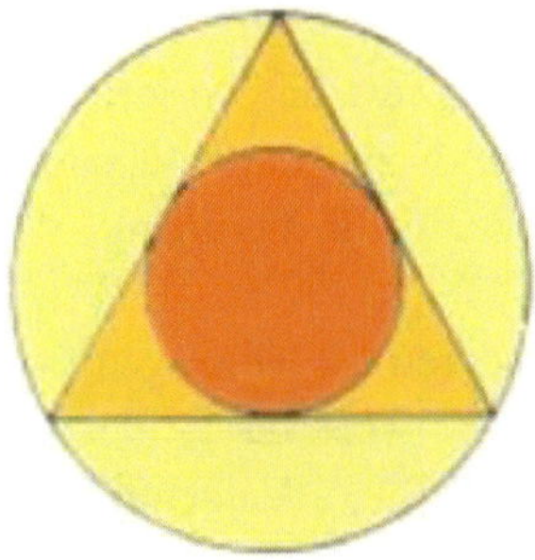

Theorem II

Figure 5. Theorem II (Diagram)

Sacred Geometry

The word Geometry itself means 'Earth measure', which definition is generally attributed to the fact that the ancient Egyptians regularly utilized geometry to resurvey the fertile farmlands of the Nile river floodplain in late summer, after existing boundaries were buried by the deposition of thick layers of alluvium from the annual flooding of the river.

Most of us tend to think of geometry consisting of axioms, definitions, postulates and proofs, hearkening back, in fact, to the methodology of Euclids elements, in form and structure an exposition of logical thinking and mental training. While the modern, academic approach to the study of geometry sees it as the very embodiment of rationalism and left brain, intellectual processes, which indeed it is, it has neglected the right brain, intuitive, artistic dimension of the subject.

Robert Lawlor addresses this fundamentally dualistic nature of geometry in his work: Sacred Geometry – Philosophy and Practice (1982), in reference to a medieval representation of geometry as a woman seated at a table, with compasses in hand, surrounded by the implements of the art. In his words [95]:

> Geometry as a contemplative practice is personified by an elegant and refined woman, for geometry functions as an intuitive, synthesizing, creative yet exact activity of mind associated with the feminine principle. But when these geometric laws come to be applied in the technology of daily life they are represented by the rational, masculine principle: contemplative geometry is transformed into practical geometry.

Ancient geometry does not rest on a priori assumptions or axioms. The starting point is not a network of intellectual abstractions, but a meditation on the metaphysical unity visually symbolizing the formal order of this incomprehensible unicity.

Sacred geometry is considered the language of creation.

It studies the connections between the proportions and forms contained in micro and macrocosm, in order to understand the unity that pervades life. Since ancient times, the Egyptians, the Greeks, the Mayans, the architects of Gothic cathedrals and artists such as Leonardo da Vinci recognized in nature special shapes and proportions that translated into harmony and unity. These relationships between shapes and proportions, considered sacred in geometry and architecture, are also found in other areas of human expression, as in music [96].

We also find many images of sacred geometry in crop circles.

In the range of crop circle designs we see recurring themes that are for the most part generated from within a circular form and continue, through proportional expansion, to develop well beyond the boundaries of the design. This is consistent with the principle of sacred geometry where the circle is the principal element since it lies at the heart of the creative principle. All things are divided from within itself so, paradoxically, all things are contained within it. It is therefore the symbol of the unknowable, of spirit and of heaven.

The symbolic opposite of the circle is the square which is considered material and of the earth. Both forms, when given equal areas and superimposed, become a symbol of the fusion between humanity and the universe, of spirit and matter.

The division of the circle into two equal size circles, with the centers of their respective circumferences touching each other, gives birth to an overlapping figure of profound symbolism: the vesica piscis. It is from within its vagina-like center that the symbols of sacred geometry are born.

From the vesica we can create the simplest of polygons, the equilateral triangle. The two triangles contained in this 'womb' are themselves symbolic of the world above and the world below, just as the left circle can represent spirit and the right one matter. It is symbolic to Christianity as representational of Jesus Christ and the Piscean age.

The overlapping of both triangles produces the hexagon, also referred to as the Star of David. The characteristics and symbolism behind this figure are immense. It is also through the natural division of a circle into six parts which allows six circles to fit exactly around the circumference of an equal seventh. Moreover the hexagon is symbolic of the order of the universe in the fact that twelve equal spheres can be placed around a thirteenth so that each touches the nucleus and four of its neighbors, producing the geometer's image of twelve disciples grouped around the master. Christ, Osiris and Mohammed are among those who are represented as a central sphere with twelve retainers. It is also a common symbol in Islamic geometry, and its qualities are rational and solar.

Another symbol of great fascination to ancient and modern geometers is the golden mean since it gives rise to a spiral form which occurs in many natural forms and patterns of plant growth.

It is from the golden mean that we also extrapolate the pentagon and the five-pointed star, and with it the association with humanity since the human figure with outstretched limbs is similar to the pentagram. This was the symbol of the Pythagorean's humanistic science and consequently worn as a talisman of good health. The pentagon is lunar; the angle between two sides of a pentagon is 108 degrees, 1080 being the characteristic lunar number.

These primary geometric forms, along with their 3-dimensional counterparts, the Platonic Solids, were thus observed by many ancient cultures to be the crystallizations of the creative thoughts of God as they emerge from the circular Unity. As a metaphor of universal order, sacred geometry is pervasive in imagery throughout history, from circular mandalas to sacred temple [97].

Returning to the crop circles:

Vesica Piscis

The vesica piscis is a shape that is the intersection of two circles with the same radius, intersecting in such a way that the center of each circle lies on the perimeter of the other. It is the expression of two-ness, as this is how geometers depict the splitting of one into two.

This geometric pattern can be found in many artistic works of antiquity.

It is a symbol that can be considered as the basis of all forms of existence in the Immanent World – the Created World, a symbolic design consisting of a vertical ovoid, with two vertices at its ends, resembling the schematic drawing of a fish, as well as the air pocket that fish have in their abdomen to facilitate floating, being its name derived from the Latin Vesica (bladder) Piscis (fish).

It is therefore a form that represents the Universe itself manifested; a shape that results from the intersection of two perfect circles.

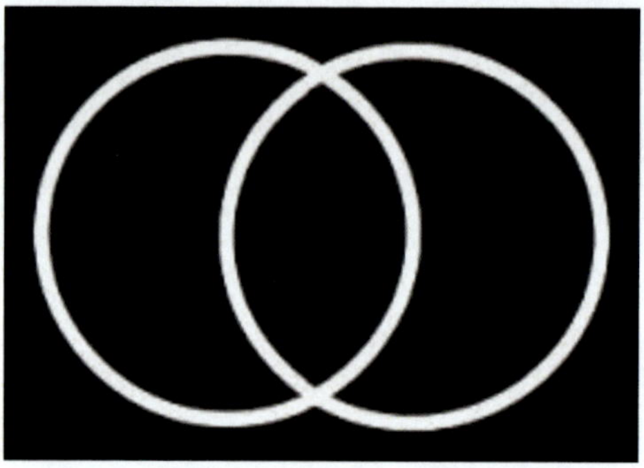

Figure 6. Vesica Piscis.

Figure 7. Ashbury, Oxfordshire, August 1st 1996.

The Flower of Life

It is a very ancient symbol found in the Vedas and also in the Celtic civilization. The Flower of Life is the pattern of creation and life, everywhere.

This is a geometrical figure composed of various circles of equal diameter, overlapping in a controlled manner, forming a structure similar to a flower composed in its core by six symmetrical petals. An infinite chain of circles that form a harmonious web within which geometric figures emerge, sacred to many ancient spiritual traditions, the center of each circle is positioned exactly on the circumference of the six circles surrounding it.

According to the sacred geometry, each molecule of life, every cell in our body knows this pattern, and all knowledge in the universe is contained in it. All harmonics of light, sound and music are in this geometric structure, which exists as a holographic pattern. The code of the Flower of Life contains all wisdom, similar to the genetic code contained in our DNA.

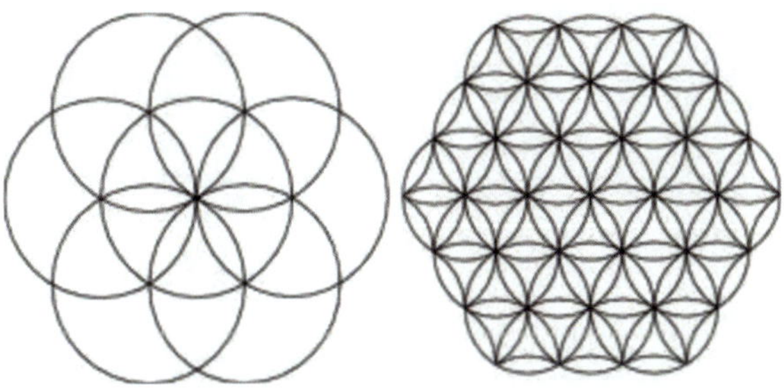

Figure 8. Flower of Life

Figure 9. Froxfield, Wiltshire, August 5th, 1994.

Plato's Polyhedra

Plato, around the sixth century BC, studied certain class of polyhedra, called regular, which later became known as Plato's polyhedra. Regular polyhedra are defined as having faces that are regular polygons, all identical, and all angles are also equal to each other. In "Timaeus", Plato creates an association of the five regular polyhedra with the five elements in nature. He considers the Tetrahedron as "source element of Fire," the Cube is associated to Earth, the Octahedron to Air, the Icosahedron to Water and the Dodecahedron would represent the image of the Universe as a whole.

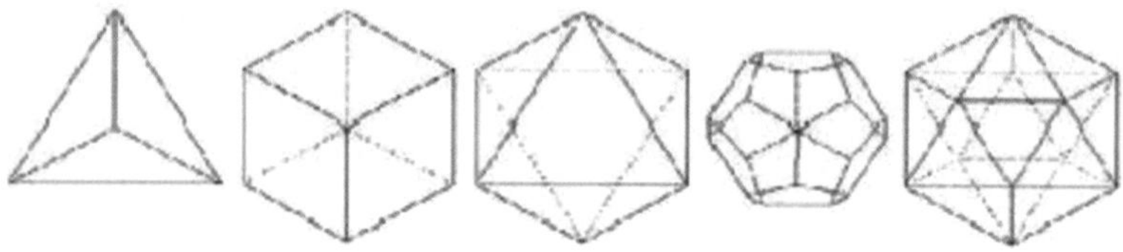

Figure 10. Platonic Solids.

Metatron Cube

The pattern of the Flower of Life is the basis of the Fruit of Life, which is, in turn, the origin of the Metatron Cube, the geometrical figure that contains within itself the two-dimensional projection of all Platonic bodies.

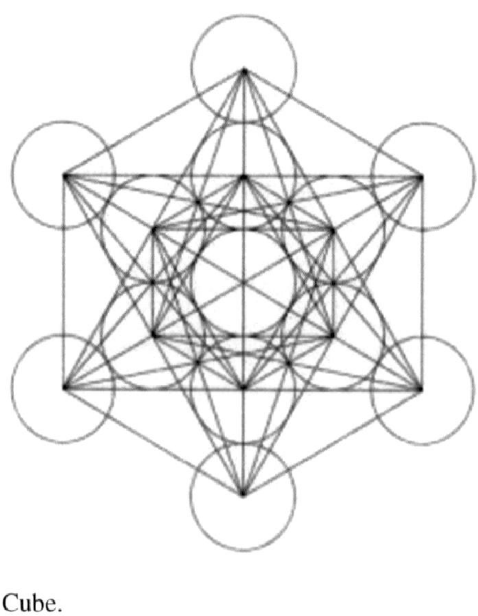

Figure 11. Metatron Cube.

The Metatron Cube is built based on the Fruit of Life, that is, 13 tangent and congruent circles derived from a regular hexagon. By joining the center of each of these circles with the centers of all the others, you get this figure comprised of 78 lines.

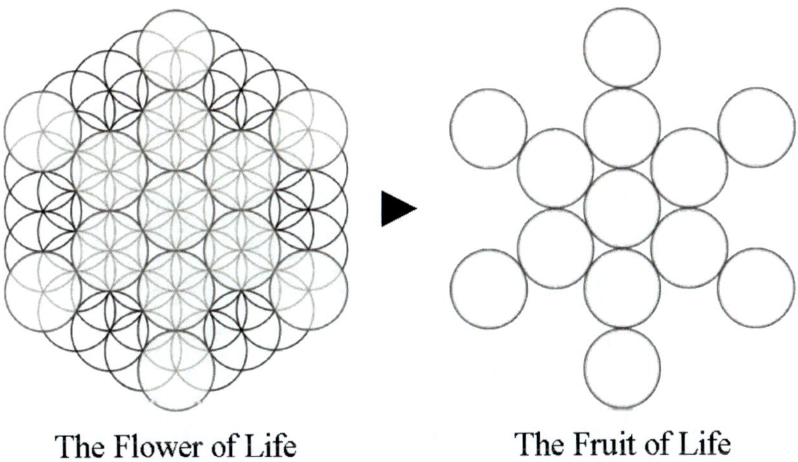

Figure 12. The Flower of Life and The Fruit of Life.

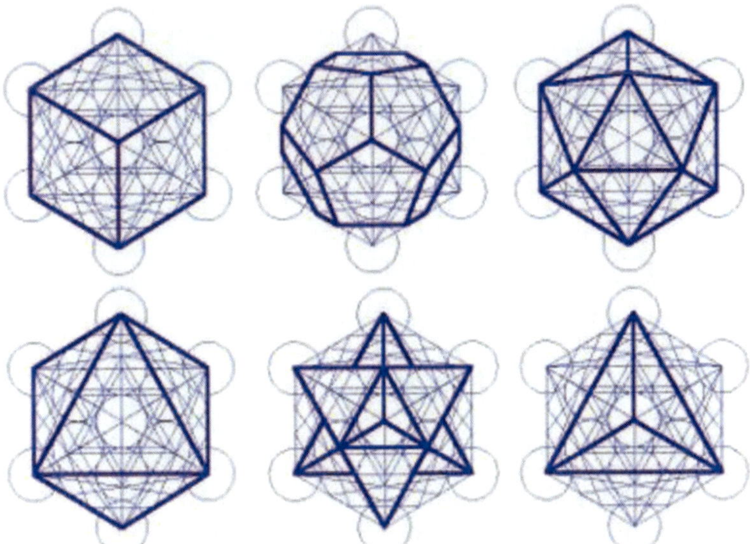

Figure 13. Metatron Cube and the Platonic Solids.

Returning to the Crop Circles

Thus, we marvel before the crop circles, astounded at the manifestation of the archetype of order in the sacred numeric and geometric patterns that arise in the fields. Having forgotten the sacred geometry and the qualitative and divine aspects of the number, we do not pay attention to its existence a priori, as a forger of the Cosmos, and previous to our achievements of expression in equations and numeric manipulation; we made of the many (infinite numbers) a magnitude, and we use them in mathematics misled by the idea that we created them.

Figure 14. Etchilhampton Hill Wiltshire, July 25th, 2011, Wheat.

We arrogantly forget that they do not belong to us, but rather come to us, just like the magnificent crop circles do.

How might we look back at those neglected aspects of numbers, sacred geometry, crop circles designs and what they embody, to the meaning that lies beneath the shapes and the perennial knowledge they seem to employ?

In Zen philosophy, they speak of something called “Beginner’s Mind”.

It is a state of ‘focused openness’ where our preconceptions and everything we think we already know about something, are temporarily set aside.

A zen tale: Empty Your Cup [98]

The story concerns a Buddhist scholar and a Zen Master. The scholar had an extensive background in Buddhist Studies and was an expert on the Nirvana Sutra. He came to study with the master and after making the customary bows, asked her to teach him Zen. Then, he began to talk about his extensive doctrinal background and rambled on and on about the many sutras he had studied.

The master listened patiently and then began to make tea. When it was ready, she poured the tea into the scholar's cup until it began to overflow and run all over the floor. The scholar saw what was happening and shouted, "Stop, stop! The cup is full; you can't get anymore in."

The master stopped pouring and said: "You are like this cup; you are full of ideas about Buddha's Way. You come and ask for teaching, but your cup is full; I can't put anything in. Before I can teach you, you'll have to empty your cup."

CONCLUSION: *UNUS MUNDUS* AND CROP CIRCLES

> "Without the reflecting consciousness of man, the world is a gigantic meaningless machine, for in our experience man is the only creature who is capable of ascertaining any meaning at all."
>
> *C. G. Jung*

We must first remember that our empirical, psychic and material reality is the manner how our polarized consciousness presents us our psychophysical transcendent dimension, on which everything is based. The phenomenon of synchronicity gives us evidence of this dimension when it erupts, in an acausal manner, carrying a coincidence of physical and psychic meanings. The appearance of crop circles would also be a manifestation of this transcendent dimension of absolute order and knowledge in our time-space physical reality, causally conditioned. As discussed earlier, the idea of *Unus Mundus* is based on the assumption that the multiplicity of the empirical world rests on the foundation of its unit. Thus, all that is separate or different belongs to a single world, which in turn does not appear to the senses, but rather represents an assumption. This transcendent dimension, the *Unus Mundus*, to which we have no access, comes to us through the symbolism of mandalas: it contains the core or the ultimate unit of all archetypes and everything around it, representing the multiplicity of the phenomenal world.

The simple circle is the ultimate expression of one-ness. Unity can be appropriately represented as a circle, but the very incommensurability of the circle indicates that this figure belongs to a level of symbols beyond reasoning and measure. A circle has no beginning and no end, representing the eternal, wholeness, unity and perfection. Another unique and mysterious property of the circle is that, unlike any other geometrical shape, the sum of its perimeter is unknowable. We therefore approximate the perimeter ($2\pi r$) since π is a number whose decimal places would stretch into infinity and is inexpressible in whole units. In the theory of The Forms, Plato postulated that we may only ever approximate a universal form, one that exists in a state of perfection beyond our physical realm since it exists in a state of flawlessness that we can only ever aspire to. In rendering a circle as faithfully as we are able, we take part in a sacred act which pays homage to the perfection of the Divine [99].

The very term 'crop circle' is derived from the fact that the phenomenon began as single, simple circles appearing in crop fields.

And thus, again, the geometry of the mandala in the fields, either erupted from the depths of the earth or printed from spiritual heights, transforms us in witnesses of a numerical and dynamic order, nondivided, transcendent and difficult to assimilate.

The alchemists' idea of a single world is that potential world of the first day of creation, when nothing existed separately. The alchemists believed that matter, arising from this potential world, possessed a divine aspect, which was completely abandoned by the scientific rational vision. Subatomic research, however, has placed science once again face-to-face with matter that does not exist by itself, leading to a relationship of mystery with the psychic. And in recent decades mainly, we have witnessed the unfathomable manifesting itself repeatedly in matter, through geometric designs in the fields. Then we realize, numinously surprised, that they correlate with the deepest psychic images, both objective and numerical, of psychoid order. This correlation again leads us to suspect that there is a common origin; this single world of which we can only grasp perspectives.

As I understand it, the origin of the phenomenon is not psychic or material, but, as said before, a manifestation of the transcendent *Unus Mundus*. And, although still not seen as part of the divine and without receiving the deserved attention, crop circles are a reality that came to make us questions, enchant and bring us new directions. They carry the meaning of unity that our dissociated time greatly needs and fears to find without being able to explain. That meaning that links us with the numinous – the meaning of the Whole, the Tao – and, so to speak, puts us once again in our proper place in the Whole. It gives us a feeling of 'that which is just as it is'; it brings a reconciliation with life and death, joy and sorrow, conflict and peace. It is not nirvana, a spiritual 'beyond', but rather a complete acceptance of "what is" [100]. They open up the divinity in our psyche and matter, trying to draw our attention to a unified psychophysical truth, in a moment so devoid of true connections. With their circles, they overshadow our square truths so that many of us still refuse to see their beauty, reducing them to a scam or hoax. And last, in my view, they demand responsibility and ethics from us humans, inflated and empty of mystery, who believe that we can still control and subjugate everything around us. There is no ethics in life prescinding from a greater meaning to which we are called and confronted with. For me, crop circles are themselves this call.

> “You cannot go on "seeing through" things forever. The whole point of seeing through something is to see something through it. It is good that the window should be transparent, because the street or garden beyond it is opaque. How if you saw through the garden too? It is no use trying to "see through" first principles. If you see through everything, then everything is transparent. But a wholly transparent world is an invisible world. To "see through" all things is the same as not to see.”
>
> *C. S. Lewis*

> “Let the crop circles awaken us: so that we flex…before the divine present on earth; and reflect…on earth, the divine of our psyche.”
>
> *(Simone Mancini Castilho)*

REFERENCES

[1] Ken Keyes Jr. "*The Hundredth Monkey*"

[2] Jung, C. G. (1946). *On The Nature of The Psyche*. C.W. 8/2. Princeton University Press. 2001. par. 700

[3] Mansfield, V. (1995). *Synchronicity, science, and soul-making*. Illinois: Chicago and LaSalle.

[4] Goswami, A. (2007). *The Self-Aware Universe: How Consciousness Creates the Material World*. Putnam's Sons, 1995.

[5] Carriérre, Jean-Claude. El circulo de los mentirosos/ The Circle of Liars: Cuentos Filosóficos Del Mundo Entero/ Philosophical Stories of the Entire World (Palabra En El Tiempo/ Word in the Time) (Spanish Edition). Lumeneditorial, 2001.

[6] Rocha Filho, J. B. (2007). Física e Psicologia: as fronteiras do conhecimento científico aproximando a Física e a Psicologia Junguiana. (Physics and Psychology: The frontiers of scientific knowledge approaching the physics and Jungian psychology) Porto Alegre: Edipucrs.

[7] Jung, C. G. (1934). *The Archetypes and the Collective Unconscious*. C.W. 9/1. Princeton University Press. 1981. par. 3

[8] Von Franz, M. L. (1992) Matter and psyche from the point of view of Jung. In: *Psyche and Matter*. Boston & London: Shambhala.

[9] Silveira, N. (1997). Jung: vida e obra. (Jung: life and work) São Paulo: Paz e Terra.

[10] Jung, C. G. (1946). *On The Nature of The Psyche*. C.W. 8/2. Princeton University Press. 2001. par. 439.

[11] Ibid.

[12] Available from < http://www.arvindguptatoys.com/arvindgupta/zen-for-neighbours.pdf>. Accessed on: June 20, 2014.

[13] Ibid.

[14] McFarlane, T. J. (2000). *Quantum Physics, Depth Psychology, and Beyond.* Available from: <http:// www.integralscience.org>. Accessed on: Oct 10, 2013.

[15] Jung, C. G. (1935). *Analytical Psychology: Its Theory & Practice (The Tavistock Lectures).* Vintage. 1970.

[16] Ibid.

[17] Ibid.

[18] Ibid.

[19] Ibid.

[20] Meier, C. A. *Atom and Archetype: The Pauli/Jung Letters* 1932-1958. London and New York: Routledge, 2001.

[21] Ibid.

[22] Magia Oriental. Available from <http://aoikuwan.com/2012/02/24/conto-zen-mente-em-movimento/>. Accessed on: November 5, 2013.

[23] Ibid.

[24] Ibid.

[25] Miguelez, M.M. (1993) El paradigma emergente: Hacia una nueva teoria de la racionalidad cientifica In: Historia de la ciencia y epistemologia (The emerging paradigm: Towards a new theory of scientific rationality. In: History of science and epistemology). Spain: Gedisa Editorial.

[26] Ibid.

[27] Ibid. par. 270

[28] Ibid. par. 417

[29] Ibid. par. 416

[30] Ibid. par. 398

[31] Ibid. par.418

[32] Ibid. par.405

[33] Ibid. par.414

[34] Ibid. par.420

[35] Aufranc, A. L. (2009). A questão do sentido no mundo do acaso. Junguiana (The question of meaning in the world of chance). São Paulo, n. 27, p. 45.

[36] Xavier, C, R. (2003). A permuta dos sábios: um estudo sobre as correspondências entre Carl Gustav Jung e Wolfgang Pauli (The study exchange of the sages: A study of the correspondence between Carl Gustav Jung and Wolfgang Pauli). São Paulo: Annablume.

[37] Jung, C. G. (1950). *Synchronicity: An Acausal Connecting Principle.* C.W. 8/3. Princeton University Press. 2010. par. 954.

[38] Ibid. par. 849
[39] Ibid. par. 830
[40] Ibid. par. 819
[41] Von Franz, M. L. (1980). *On Divination and Synchronicity: The Psychology of Meaningful Chance*. Inner City Books.
[42] Ibid. par. 955
[43] Ibid. par. 938
[44] Ibid. par. 952
[45] Ibid.
[46] Ibid.
[47] Von Franz, M. L. (1974) *Number and Time: reflections leading toward a unification of depth psychology and physics*. Evanston: Northwestern University Press.
[48] Ibid.
[49] Jung, C. G. (1956). *Mysterium Coniunctionis*. C.W. 14/2. Princeton University Press. 1977. par. 422
[50] Ibid.
[51] Ibid.
[52] Ibid.
[53] Von Franz, M. L. (1974) *Number and Time: reflections leading toward a unification of depth psychology and physics*. Evanston: Northwestern University Press.
[54] Ibid.
[55] Ibid.
[56] Ibid.
[57] Ibid.
[58] Ibid.
[59] Ibid.
[60] Ibid.
[61] Lawlor, R. (1982). *Sacred Geometry: Philosophy & Practice*. Thames & Hudson.
[62] Ibid.
[63] Ibid.
[64] Ibid.
[65] Ibid.
[66] Ibid.
[67] Ibid.
[68] Ibid.

[69] Glickman, M. (2009). *Crop Circles: The Bones of God*. England: Frog Books.

[70] Available from: <http://www.sheldrake.org/files/pdfs/Cropcircles_Michellany.pdf>. Accessed on: June 20, 2014.

[71] Ibid.

[72] BLT Research Team Inc. Available from: <http://www.bltresearch.com>. Accessed on: October 10, 2013.

[73] Available from: <http://www.cropcirclesecrets.org/biophysical.html>. Accessed on: June 20, 2014.

[74] Ibid.

[75] Ibid.

[76] Haselhoff, E. H. (2001). *The deepening complexity of crop circles: scientific research and urban legends. Berkeley*, CA: Frog, Ltd.

[77] Ibid.

[78] Ibid.

[79] Ibid.

[80] Jung, C. G. (1976). Visions Seminar. USA: Princeton, 1997.

[81] Ibid.

[82] Bobroff, G. S. (2011). *Jung, crop circles & the reemergence of the archetypal feminine*.

[83] Jung, C. G. (1968). Man and His Symbols. Dell.

[84] Alexander, S.; Alexander, K. (2006). *Crop Circles: Signs, Wonders & Mysteries*. London: Arcturus Publishing Limited.

[85] Ibid.

[86] Jung, C. G. (1939). *Psychology and Religion*: West and East. C.W. 11. Princeton University Press. 1975. par. 92

[87] Jung, C. G. (1959). *Flying Saucers : A Modern Myth of Things Seen in the Skies*. C.W. 10/4. Princeton University Press. 1979. par. 595-625

[88] Ibid. par. 625

[89] Ibid. par. 608

[90] Ibid.

[91] Ibid.

[92] Ibid.

[93] Shesso, R. (2007). *Math for Mystics: From the Fibonacci sequence to Luna's Labyrinth to the Golden Section and Other Secrets of Sacred Geometry*. Weiser Books.

[94] Ibid.

[95] Ibid.

[96] Ibid.

[97] Available from: <http://www.cropcirclesecrets.org/crop_circles_sacredgeo2.html>. Accessed on: June 20, 2014.

[98] Available from: <http://www.prairiewindzen.org/emptying_your_cup.html>. Accessed on: June 20, 2014.

[99] Ibid.

[100] Ibid.

AUTHOR CONTACT INFORMATION

Dr. Simone Mancini Castilho
Psychiatrist and Trainee Analyst
(Sociedade Brasileira de Psicologia Analítica)
Master's Degree in Medicine University of São Paulo,
Rua Lírio do Amazonas, 20
Jardim Haras Bela Vista
Vargem Grande Paulista
06730-000 Brazil
Tel: 55-11-994236099
E-mail: simonemcastilho@hotmail.com

INDEX

A

B

C

D

E

F

G

H

I

P

Q

R

S

T

U

V

W